Grantley F. Berkeley

The English Sportsman in the Western Praries

Grantley F. Berkeley

The English Sportsman in the Western Praries

ISBN/EAN: 9783742898647

Manufactured in Europe, USA, Canada, Australia, Japa

Cover: Foto ©berggeist007 / pixelio.de

Manufactured and distributed by brebook publishing software
(www.brebook.com)

Grantley F. Berkeley

The English Sportsman in the Western Praries

THE

ENGLISH SPORTSMAN

IN THE

WESTERN PRAIRIES.

Printed by]

THE VANQUISHED FOE. (Page 293).

[Spottiswoode & Co.

THE ENGLISH SPORTSMAN

IN THE

WESTERN PRAIRIES.

BY THE

HON. GRANTLEY F. BERKELEY,

AUTHOR OF

"LIFE IN THE FORESTS OF FRANCE,"

ETC. ETC.

LONDON:

HURST AND BLACKETT, PUBLISHERS,

SUCCESSORS TO HENRY COLBURN,

13, GREAT MARLBOROUGH STREET.

1861.

TO

THE MARQUIS OF BREADALBANE

THESE PAGES ARE DEDICATED

BY HIS SINCERE FRIEND

THE AUTHOR.

CONTENTS.

CHAPTER I.

CHAP. II.

CHAP. III.

CHAP. IV.

CHAP. V.

CHAP. VI.

CHAP. VII.

CHAP. VIII.

CHAP. XII.

CHAP. XIII.

CHAP. XIV.

CHAP. XV.

CHAP. XVI.

CHAP. XVII.

CHAP. XVIII.

CHAP. XIX.

CHAP. XX.

CHAP. XXI.

CHAP. XXII.

ILLUSTRATIONS.

THE ENGLISH SPORTSMAN

IN THE

WESTERN PRAIRIES.

CHAPTER I.

DEPARTURE FROM ENGLAND—OBJECTS OF TRAVEL—ANTICIPATIONS—
MY GUNS AND DOGS—DIFFICULTIES AS TO THE LATTER REMOVED
BY SIR SAMUEL CUNARD—AMERICAN IDENTIFICATION—FAREWELL
AT LORD SEFTON'S—ON BOARD THE AFRICA—THE INQUISITIVE
PORTER—HOME-SICK RECOLLECTIONS—SIR JOHN RENNIE AND MR
BROWN—A TRAVELLING COMFORT HIGHLY RECOMMENDED—MY FIRST
WHALE—SHIP'S MUSIC NO LULLABY—YES, SIR!—SETTLING A MED-
DLESOME OLD GENTLEMAN—LOVELY AURORA BOREALIS—BRUTUS
AND THE FEMALE HAND—THE AMERICAN PILOT—ON SHORE.

In the summer of 1859 I arranged my plans for prosecut-
ing a mission to the prairies *viâ* New York, in search of
game, and with a view to collect as much information as
possible in regard to that gigantic country of America as
the short space of three months would permit.

When it was known among my friends in England that
I proposed to leave Liverpool on the 20th of August,
visit the plains, and hunt the buffalo, or, more correctly

speaking, the bison, in his native wilds, and return home for a Christmas dinner, the impossibility of the accomplishment of the feat was widely asserted. Some gentlemen assured me that I must pass a dismal winter at one of the advanced forts on the prairies, while others asserted that I should not reach the "buffalo grass" before the winter had closed in; others insisted that, altogether, I went at the wrong season of the year. With open ears I listened to everything that everybody had to say, while, with open eyes, I watched and had watched the length of the railways in the United States, and perceived that every year the rail progressed on that go-ahead system so natural to the nation, as the veins necessary to carry out the circulation of a mighty trade. One of my objects was to show to the rich and rising, and, I hope, adventurous and hardy, sportsmen of the present day, that, even if their duties called them to the House of Peers or Commons, they might attend the session of Parliament, visit their brother sportsmen in America, reach the haunts of and hunt the larger game, add to those they had in England the hearty friendship of many a transatlantic gentleman, and return again in ample time to share in or maintain the usual hospitalities of Christmas. That the sportsmen of England may do so, is proved by my having sailed from Liverpool in the " Africa," on the 20th of August, 1859, and returned to the same city in the " Asia," on Sunday night, the 4th of December, with the huge bison's head to crown the trophies in my hall.

Before starting on this adventure, I did all in my power to ascertain the best mode of proceeding to the prairies of the Far West, and the probable cost of, and the best outfit for, the journey. But, alas! all the gentlemen to whom I spoke, and who had been in

America, could tell me not a jot more than I knew myself as to what I should do or have to do. They said that " there were buffaloes on the plains, wildfowl in the lakes and rivers, and feathered game upon the lands," and that these " could be killed by pulling triggers ; " but how they could be reached, and at what cost, never came within the scope of their information. . In my thirst for intelligence of this sort I found a work, wherein the author assured his readers that he had passed a delightful, romantic, and even graceful life with a tribe of Indians rejoicing in the name of Pawnees, and partaken with satisfaction of the delicacies of reeking livers torn raw from the bosoms of buffaloes ; on reading which (with the exception of having no desire for the food alluded to) I had dreams of attaching myself to the Pawnee tribe, and of falling in love with the chief's daughter, and wedding her for a time by the religious and august cere- mony of tying a mule up to her father's hut. I say that I had dreams of this, but not the slightest desire that those dreams should be fulfilled. How I read of these Pawnees, and what I found the Pawnees really to be, must remain to be told in other portions of my narrative. Some gentlemen said they had killed the grizzly bear by " stalking him ; " others cried " that shooting buffalo was nothing better nor more wild than walking up to and shooting oxen in a farmyard." Whether I found this bovine assertion to be correct or not my readers will ascertain when the buffalo or bison hunt comes before them, in its double view of the "still hunt" or stalk, and " the run on horseback." In short, I found many gentle- men who, having, according to their account of the matter, done everything, had forgotten to take notice of the method, cost, and practice, the all in all as to future usefulness, by which their success had been achieved. I left England, then, not only for my own pleasure, but to be able to tell Englishmen in what way they should proceed on a visit to the Far West, what adventures they would most likely encounter, what would be the costs of their journey, and how they had best fit themselves out, and at the least expense. Due regard to personal safety also being kept in view, when they formed their camp or caravan to travel hundreds of miles over a desert, where fever and ague were rife, water scarce, roving robbers of the white class many, and Indians in thousands, whose readiness for war-paint or peace could never for an instant be relied on.

Besides ascertaining these things, so requisite to pleasure, sport, and the gaining of useful knowledge—for a man cannot reap the benefit of wide research unless his mind is free to general and uninterrupted application—I had it in contemplation to ascertain which of the English dogs, or how many of their kinds, might be made serviceable in America. Of the climate, in regard to English dogs, I had no fear; the sorts of game, and the nature of the country over which that game had to be sought, were the points on which I deemed it necessary to obtain information. Another question also had to be set at rest, and that was the species of firearm best adapted to the different kinds of game. Of guns and rifles (breech and muzzle-loaders) I had great choice, and the perfection of the English make. The London gunmakers were John Manton, of Dover-street; Prince, of London; and, in addition, Pape, of Newcastle-upon-Tyne. How I applied the perfect weapons they supplied me with, and what their firearms achieved in my hands, I must also leave to the due course of the following narrative. That narrative will consist of my line of travel through the United

States, my sporting adventures, and my remarks on the political and social state of America, its society and manners. All matters in short that time, and, I hope, acute observation, brought within the notice of a sportsman in his search for knowledge.

When my mission to America was proposed to me I had plenty to amuse myself with in England; but there was a fact which occurred to my mind that met with no denial, and that was, that I should never be more fit to undertake such adventures than I was at the present moment. The only thing that caused me some uneasiness in regard to companionship was as to my favourite dogs; for, ignorant as everybody in England seemed to be as to the sort of dogs that would be useful, and uncertain as I was as to the convenience that would be afforded to them in travel, at sea and through the United States, I hesitated to put them in a position in which their safety would not be within my control.

Many friends however advised me to take them, and at last, through the kind and liberal attention of Sir Samuel Cunard, all difficulty in their transit by sea to New York was overcome, for kennels were fitted up for them on the deck of the "Africa," and they were permitted to accompany me free of all charge—a liberality and attention in no way imitated, or apparently thought of, by the railway or steam companies on the other side the Atlantic. I confess myself to have been surprised at this, for my brother sportsmen in America had been apprised of my coming visit; and as an Englishman was about to throw himself, in that sort of off-hand manner generally so pleasing to men who make the plains and forests the scenes of hearty companionship, on their generosity, I naturally expected that they would give me some sort of facility or counsel, not only as to my journey to the Far West, but more particularly as to the conveyance of my dogs, unprotected as they were by their laws, and not even recognized on the rail as things for which the company had provided any means of conveyance. When I assure my readers that this apparent neglect "surprised me," that surprise only lasted till I became better and locally acquainted with the men among whom I was about to travel. When I made their acquaintance I understood it all, and for brevity's sake, as well as perspicuity, my knowledge is summed up in the following few words:

No man who signs himself of whatever rank is generally believed to be the person he represents himself to be until some other man in America introduces him according to the signature he has given out. The President is not believed to be the President, and Mr Cobden is not believed to be the free-trade orator, when on travel, unless, like sharks, they have a pilot fish or "man in black" to verify or proclaim their individuality—hence no one believed in me.

It was a giant task that which I had to do in three months. Agriculturally speaking, a huge stubble lay before me, into which I was to go as a gleaner—a gleaner never to stoop nor stop, but to pick up everything at a run, and to reach the haunts of the larger game—the buffalo or bison to be the extreme magnet of attraction.

My bloodhound Druid and myself, Brutus and Alice my retrievers, Chance my setter, and my deer-lurcher Bar, directly after my visit to the mountains of Lord Breadalbane, set off from Beacon Lodge on Thursday, the 18th of August, 1859, and reached London that evening, where we dined

and slept. On the following morning we set off for
Croxteth (Lord Sefton's), where we dined and slept;
and on Saturday, the 20th of August, we—myself, my
dogs, and their attendant (George Bromfield)—found
ourselves at noon safely on board the " Africa " steam-
packet, of the Cunard line, of 2500 tons, bound for New
York, and under the command of that excellent seaman
and gentleman, Capt. Shannon. At the quay there were
the usual number of officious porters, all attempting to
seize some article of luggage, but by dint of extraordinary exertions of legs, arms, and tongue, I cut their
legion down to three, that number of individuals being
two more than I had any occasion for.

My luggage, then, all on board the tender, an officiat-
ing and officious man laid his hand on a deal box in which
some of my guns were packed, and to which a card was
nailed specifying of what the package consisted, and demanded payment or freightage for that lot. To this demand I returned a flat refusal, saying that the guns were
a part of my personal baggage for use, and not for sale,
and that I would not pay for them. He then asked,
" What could I do with so many guns ?—a man could
only shoot with one." I replied, " That proved *his* ignorance, for just as a man might put on a thick or a thin
coat according to the weather, I took into my hand
a different weapon according to the size of the game.
He might just as well charge a carpenter for his tools or
chisels as charge me for my guns, so I would certainly
not pay one farthing." By this time he had had opportunity to spell my name, and, apparently laughing at
the idea of the carpenter, he desisted from any further
attempt to move me out of money.

On reaching the packet I found it immensely full, the

living freight comprised of people from almost every
clime under the sun, with a large sprinkling of the natives of that nation to whose wide realms I was about to
pay this visit; but having secured a state cabin to myself, free from all intrusion, as well as (through Sir
Samuel Cunard's great kindness) comfortable cages for
my dogs, whence they could annoy no one, nor be annoyed, I made myself as contented as a man could be
who might be said to have left the dearest and best behind him, and the site of all his happiest hours.

However new and fresh the scene, and however much
I might delight in adventure, still it was impossible to
drive from my soul the knowledge that every revolution
of the wheel took me further from those I loved, and
shut me out for a time from the graceful enjoyments of
life, which can nowhere be found so perfect as in England. Even on the broad bosom of the Atlantic, "the
glad waters of the dark blue sea," with the long glittering feather the ship left behind her, failed to cheer me,
and a thousand recollections by night and day crowded
on my mind, to the exclusion of anticipations of pleasure, and at first I found it impossible to fasten observation on anything but the lapse of the last English hours.
Time, however, that best physician for all human ills,
who, differing from the mortal faculty, so frequently
cures before he kills, came to my aid, when, though his
skill could neither banish remembrance nor stifle regret,
nevertheless he called to his aid perhaps the sunbeams
glittering on the sea, and through health, strength, and
the elasticity of life, he coaxed me to feel that I had only
banished myself for a time, and that under the blessing
of Heaven I should soon return in full recollection and
fondness for all, and to ask if I had been forgotten, and

if no change had come over a spirit, or *the* spirit, of a dream, that seemed to go with me as my shadow?

The "Africa" is an excellent sea-going, safe ship, like her twin sister, the "Asia," in the Cunard line, and if not a very fast one, still she is very sure. In the tender or boat which took me on board her from the quay at Liverpool, I renewed an acquaintance with an old hunting friend whom I had not seen for years (Mr Brown), and made the acquaintance of Sir John Rennie, both bound for New York, but not on missions like my own. In the society of these two friends I had much amusement, though Sir John insisted that I should be scalped by the Indians, and we enjoyed a vast deal of merriment occasioned by the circumstances of the hour.

The arms that I took with me to the plains were my two old favourite double-shot guns, of eleven gauge, made by John Manton, of Dover-street renown, and which I had shot with for thirty years, and my old favourite single rifle, of a similar period, from the same master hand. To these were added a double-shot gun, most splendidly put out of hand by Mr Pape, of Newcastle-upon-Tyne, slightly less in gauge than eleven, and a double breech-loading rifle, with the conical ball, by the same maker—also a first-rate article, needing no additional cap, the cap being contained in the cartridge; a breech-loading carbine by Mr Prince, of London, on the pattern of that arm for cavalry, but made to carry a heavier ball, for the benefit of the buffalo or bison, to which special service I intended to address it. These, with a powerful double rifle made by Collins, lent me with a revolver by Captain Bathurst, of the Grenadier Guards, and a splendid East Indian dirk, or *couteau de chasse*, kindly put into my hands by Colonel

Brown, made my arm-chest an ample one, and fitted
me out with every sort of weapon or firearm that sport
or danger might require. In alluding to my weapons
I must not forget a perfect long hunting or clasp knife,
which had been given to me, containing, besides its
trenchant blade, a lancet, tweezers, and cork-screw, all
very useful things in forest, field, or camp.

The first thing that a traveller ought to do on board
a packet is to look to the due arrangement of his luggage, and before it is all stowed away in the " hold "
assigned it, to point out to the steward that part of it
which is necessary to his personal comfort on the voyage, and which can be put in his private cabin. As an
article of the greatest use in carrying clothes, as well
as for its extreme comfort in other respects, let me
advise every traveller to come, whether male or female,
to provide for themselves an oval-shaped tin bath, or
packing-case bath, with strap and lock. I procured one
from Mr Barrett, at the corner of Albemarle-street, in
Piccadilly, and it not only gave me the fresh sea wave
of the Atlantic every morning throughout my voyages
for my bath, but it was the only packing-case I had
that completely withstood the brutal and dishonest usage
of the baggage-masters on the American railways, as
well as the rough but fair wear and tear of waggon-
carriage during my camping out on the plains of the
Far West. This sort of package not being universal,
or at least much delighted in by travellers in the United
States, I shrewdly suspect that the men in the baggage-
cars on the railway regarded my tin bath as a magazine
of gunpowder or combustible matter, and not as a package
containing anything they could steal, if by knocking it
intentionally about they could start the lock. Hence,

for their own personal safety, as well as from the non-expectation of gain, I believe arose the immaculate state in which it returned to England.

Having put my things in order in my berth, and forced my eyes from their internal retrospection, I knew that we had left Liverpool at 1.45 p.m., and had discharged the pilot at about 3 p.m. off the Bell Buoy. Steaming by the North Channel, we passed between the Calf of Man and the Chickens, South Rock abreast of us on the Irish coast. At 4.10 a.m. on Sunday we were abreast of the Maiden's Lights; at 6.5 a.m. entered the Sound of Mahu; at 10 a.m., in the Sound of Innistrahull, the extreme north point of Ireland; the Tory lighthouse was then the last land we saw. We then sped on our way with fine weather, but with a head wind and swell which sent innumerable passengers to the privacy of their berths. Having been some days at sea, the first living thing that interested me on the broad bosom of the Atlantic was a whale on our starboard quarter; by the thermometer the temperature of the air 66°, water 60°. The next, an ornithological observation, was the fact of a flock of gulls, not as birds of passage, but occupied in fishing, and at the time I saw them we were from six to seven hundred miles off the coast of Greenland, the nearest sea-shore to where we were. I make this remark because I have heard it stated that the common gull is seldom seen so very far out at sea.

On the 27th we had thick weather, a false horizon at noon, to be guarded against in taking observations, and plenty of Mother Carey's chickens playing around us, their flight along the crest of the waves resembling that of the swallow. Fog cleared away in the afternoon,

and rising, like the gauze curtain of a theatre, gave
us the distant but beautiful and terribly-interesting view
of a majestic iceberg on the port-bow, its distance from
us not correctly to be computed, but guessed at about
ten miles, and its length about three hundred feet by
ninety. The size and pallid yet colossal appearance
of this spectre on the deep made me perfectly aware of
the danger of collision in those dense fogs that frequently cumber these latitudes, and pray for clear weather.
With night, however, the fog returned; we were
reduced, for the purpose of greater safety, to half speed,
and there was no sleeping on account of the fog-whistle
sounding at intervals throughout the night. On the
second Sunday after leaving Liverpool some of us
expected to sight Newfoundland, for the weather had
again become clear, fine, and calm, but in this expectation we were in error. During the voyage the weather had been, in my opinion, very fine, from first to
last, with two whales, the iceberg, and a very large
shoal of very small porpoises, gulls, divers, and Mother
Carey's chickens to amuse me; but among my fellow-
passengers, and on account of the head wind and swell
and long pitch of the vessel, there was a vast deal of
sickness. One old lady, I regret to say, died from the
effects of sea-sickness three days prior to reaching New
York, and it was deemed proper to carry the body to
the United States.

Throughout this voyage, then, we had fine sea-going
weather, with two or three fresh breezes at night, which
rendered pillows far from restful, and occasionally sent
undressing victims reeling from side to side, cannoning,
as a billiard-player would express it, against every corner
of their berths, and in all probability pocketing them-

selves in their wash-basins. Fortunately for me, though
at times I have felt *odd*, I am never sea-sick, but in full
possession of my senses, so that I could observe all that
passed, and read, and even write, and, as Capt. Cuttle
would say, "make a note of it;" hence nothing was
lost to me, and in their lucid intervals I had opportunities of
much conversation with the men of the country in whose
land I was so anxious to arrive—anxious indeed, for, of
all places that so soon pall upon the passenger's mind,
there is none that gets so wearisome as the unvaried
confinement of a ship at sea. At night the same cramped
crib to lie in, never long enough for a man of more than
six feet one, and therefore too short for me, and always
too cold or too hot. The same noises góing on in the
timbers of the vessel and panels of your berth ; in mine,
in the "Africa," it was as if all the rats and mice on earth
had distressfully combined with innumerable sucking-pigs
and an old sow or two, and some ducks, aided by the
constant knocking of little hammers variously placed, to
raise dins against going to sleep ; and when I left my
berth a horrible admixture of incongruous smells assailed
me from kitchen, crevice, chimney, cabin, and cockpit, and
it was only on deck that I could catch the sweets of untram-
melled air. Every person, of whatever degree, from the
United States, seemed pleased to make my acquaintance,
and proffer me the hand of good-fellowship. "Yes, sir;"
I can't help using these two words in confirmation of this,
though they—some of our cousins over the water—
use them so repeatedly and oddly when they mean nothing confirmatory at all.

There was scarce one of these gentlemen that did not
give me a hearty invitation to his house, and press me to
come there, many of them adding, "You can't mistake

my house, as my name is up in large letters over the
door. Yes, sir." During the voyage my dogs were an
object of interest to all, and I was requested to let them
come in fine weather to the upper deck; but I did not do
so, for had a lady's glove or handkerchief blown over the
side of the ship, dear Brutus would have leaped after it,
and Druid is never safe among strangers. With the
sailors my dogs were such favourites that no orders I
could enforce would prevent their getting too fat; every
jolly fellow would give them a "bit of his grub" whenever he had any himself, and this with confinement
militated much against condition. The dogs were all
sea-sick, and so was the ship's cow; so no disgrace to the
sea-going stomachs of my kennel! The only occasion on
which I had cause to be angry during the voyage out
was when from the upper deck I heard all my dogs rav-
ing with fury, Druid's and Brutus's voices being of course
predominant. I knew that my servant George suffered
from the sea, so my attention was always alive to the
things he had under his charge, but could not duly look
to. On going forward, to my astonishment, I found an
elderly cabin passenger, grey enough to have known better, stirring my favourites up with a stick, and amusing
himself with their fury at not being able to punish him.
"Ho, sir!" I exclaimed, as I arrived unexpectedly at
his elbow, "don't you think there's danger in thus teas-
ing my inoffensive dogs, with the deep sea so near?"
He coloured, said nothing, and walked away, and I never
had further occasion to notice him, or complain of such
wanton aggression.

I had not been long in the society of gentlemen from
the United States before I learned that to tell untruths
was not to lie. To tell an American "he lies," is to

bid him draw a knife or revolver ; but pleasingly to show
him that you guess " he's pitching it in considerable
smart," and departing from unsmart fact, is no insult
whatever. " Do you shoot with rifle ? " said one of my
conversationally playful friends to me. " Yes," I replied.
" Guess you think then you're a pretty good shot," he continued. " Guess I do," was the rejoinder. " Guess you'll
not come up to our Kentucky men," continued my companion ; " better not try your hand there unless you can come
nigh their doings." " What do they do ? " I inquired.
" Just this, yes, sir ; they place an old pea rifle barrel
horizontally at one hundred yards, and then with their
other rifle fill up the small barrel with bullets without
missing a shot, I reckon ; yes, sir ! " " I can't do that,"
I replied ; " yet I have not the least fear but that at the
living thing they will not get far ahead of me." " Reckon
we shall see," and my friend then whistled and walked
the upper deck.

Among the beautiful things I saw during this voyage
were several unclouded sunsets, and when the sea is
smooth, and the ship steady, I know scarcely anything
more sublime than the view of that splendid globe of living
fire sinking into the sea. On the ocean nothing impedes
the view. The slanting sunbeam marks, as it were, a glittering path upon the rejoicing waves, by which the lord
of light descends, and when the green and gold no longer
flash on the mighty mirror, a crimson blush succeeds,
in which effulgence all too quickly dies. We were fortunate enough also to witness perhaps as magnificent an
aurora borealis as the eyes of man ever beheld. Night
had closed in, when suddenly a new or sort of second
twilight came. The heavens at first were pale, then pink,
and green, and pale, and rosy, the tints alternating and

mingling one with the other, till at last there seemed in
the central sky, immediately over the ship, the top or spot
whence a bell-like tent of varied light began, and its
delicately-tinted walls fell all around us, even to the
horizon's verge. This aurora borealis was seen and remarked on all over America; and to me nothing could be
more lovely.

One darling incident occurred in regard to my retriever
Brutus, which must not be passed over. All my dogs
were, of course, in their more gentle feelings, outraged
and distressed by confinement; they wondered, naturally
enough, why they were, on board the ship, denied that
association with me which they had in their own nice
home. On a fine day during the voyage, when all the
passengers were resuscitated and in possession of the due
application of their legs to the motionless deck of our
smoothly-going ship, a very nice young lady from the
United States took my arm to pay a visit to my four-
footed companions, and she had on a coloured dress,
whose hue I was doomed to see remembered by Brutus
as well as myself. When we approached, Brutus was lying like a sphinx, with his head between his fore arms, and
his nose between the bars of his cage or kennel. The
instant he caught sight of the young lady on my arm
there was no mistaking whom he took her for. The soul
by man denied to the dog was in his eyes, and in affection and expectation to be set at large, trembling all over,
he met the hand the young lady put to the bars that confined him,—but what a revulsion in his manner! He, on
the instant, knew it was not the hand of happier hours,
and, with a growl, he drew back to the rear of his hutch,
while my pretty companion asked me *why* he was so

cross. I did not tell her ; and then, having inspected the other dogs, we resumed the upper deck.

I scarcely know if I ever saw a more graceful thing at sea, the wind blowing fresh, than the advent of the American pilot on board to take us into New York. With all sail set, her fore-foot often clean out of water, and lying down to every fresh squall that sent her, like a willing steed, leaping over each succeeding wave, she came, giving us an occasional view of every inner plank she had, and of the pilots as they lolled all motionless about her, in very easy and, therefore, graceful positions—not a limb stirring save when the hand at the helm gave the boat her lively duty, and at last brought her up, all shaking, on the leeward quarter of the packet. They did this so well (they knew they did it well), and the sight was so pretty, that I joined in a cheer to them from several of my brother passengers.

Early on a bright fresh morning, on the 3rd of September, I was aroused from sleep by my friends either sending to me, or calling to me, that the coast of the United States was visible. I got up, and, in the clear morning sky, saw the blue outline of a still distant land, and, what to me was also amusing, a bird of the lark species—precisely like the English bunting-lark, but called in America, so far as I could unravel their nomenclature, which is far from being correctly descriptive, the " shore lark "— come aboard of us wet and tired. On and on we sped, the land looming every moment more distinct, with a light-house visible, till gradually, but charmingly, Long Island on the starboard, and Sandyhook, Navesink, and Staten Island on the port bow, jointly with the shoaling sea, introduced us to the magnificent harbour of New York.

CHAPTER II.

In our way up the Bay of New York, merchant ves-
sels, small steamers, and what in England would be
called " coasters," were steering in all directions, and
even these, with comparatively speaking the colossal
bows of an English ship above them, and water enough
to annihilate below, in the event of collision, every sail
and wheel of them, seemed to illustrate a democratically
childish desire at the risk of self-destruction to proclaim,
" I'm as good as you, Master Englisher, we're all equal
here, yas sir ; so we reckon, Capting Shannon, those
two thousand five hundred tons o' your'n shan't make
our half-deckers get out o' the way, no how, yas sir ! "
Once or twice Capt. Shannon's muttered but suppressed broadside within his own waistcoat, amused
me, when a lolling " Boh-hoy " at the tiller would not

give even the turn of his wrist for the safety of his own life and that of his vessel, but made the steamer absolutely veer from her just course to avoid the odium of very properly destroying an ass who thought it beneath him to care for his own existence. Every revolution of the steamer's wheels, or, in other words, everything I saw, made me begin to suspect that, instead of being ushered into a land of appreciable and beneficial liberty, I was about to see the worst of all slavery, that beneath the feet of a democracy, or where millions of kings were more easily to be found than one legally ruled subject, or than a generally industrious man. An impossible classification struggling for maintenance, the component parts of the societies, at the present time sought to be amalgamated, as widely unfit to be " check and jowl together" as a sweep would be, black from a sooty chimney, to sit by the snowy garments of a girl dressed in all the purity of the wedding robe.

As we sped on our cautious run up the harbour, hill on hill, but not of any great height, continued to rise upon each other—shore on shore, and house on house, all beneath the influence of that clear blue sky, the air of which, in the vicinity of New York, is to me so soft and yet so bracing. Prettily situated merchants' villas, or what appeared to me to be so, or houses to which opulent men of business might conveniently and comfortably retire from the routine affairs of counting-houses, were dotted about above the bay, commanding the prettiest sea views; but there was no " Appledurcombe," or large, well-timbered estate or commanding mansion, with an extensive private acreage around it, such as may be seen in the Isle of Wight, and on all the beautiful spots on the coast of Old England. My first impression, then, on steaming up the bay, and in my first view of New York, was the same that strengthened on my mind during my stay in the United States. There was around me a magnificent bay filled with traffic, but scant of ships of war, and, as I still think, of forts on which to rely in defensive difficulties. I saw, in fact, in steaming up that bay, the significant sign of a nation whose inordinate desire for " going ahead," as well as necessity for doing so, rendered her dangerously neglectful of the defence in the rear, as well as of the lives of those citizens who, like the pools left by a swollen and headlong river, were to rise as they could, and follow on some other flood.

The good ship the "Africa" at length steamed up to the quay at New Jersey, on the contrary side the river from New York, and the passengers and the luggage, live or dead, so heavy was the freight, nearly filled the floor of the ample Custom-house shed appropriated to that especial purpose. Before leaving the ship I had arranged that my servant and dogs should remain on board, through the kindness of Capt. Shannon, and thus be safe from American dog-stealers, to restrain whom there is no other law than the private revolver, though you are liable to fine if that revolver is found in your possession. My dogs could then take their exercise within the locked gates of the dockyard. My luggage, consisting of eight or ten packages, was, with a view to anything but hasty collection, thrown broadcast from one end of the building to the other. Having, after vast trouble, and with the kind assistance of the medical officer of the packet, got my things together, I then obtained the agreeable notice of a Custom-house officer, and had to cut the cords of, and break open, every single thing that was not simply secured by lock and key. Oh, passengers to come! do not nail up and cord your boxes, or temper may tempt you to say things that had better be left alone! The Custom-house official was, however, very civil, and I got through my examination without much difficulty. Some remarks fell from him as to the number of my guns; but at last a comfortable fly or hack carriage, much smarter than those that ply for hire in England, received my property, and I cast myself into it with a desire to be driven as quickly as possible to the Clarendon Hotel, New York.

We had not progressed far, when the driver pulled up, and dismounting from his box and opening a window, most civilly presented me with one of the morning papers. "Uncommonly attentive," I muttered; "this man is an exception to the general rule of his class in the United States," and I commenced to read the news. I was very soon recalled to the knowledge that my horse was no longer in action, for, on hearing a whistle, and, looking from the window, I perceived my civil friend operatically indulging in idleness, from which I tartly aroused him by a rather angry question, of "Why the devil we didn't go on?" "We *are* going on," he replied; "we're half over the river." On this unexpected announcement (for all had been so smooth and quiet that I had no idea we had driven into a steamer), I got out of my carriage and went into the bows of the great broad-bottomed ferry-boat, to enjoy a view of the approach to the city. The prospect on that sunny day was a very pretty and a bustling one, an amusing one, from a variety of common-place things, rather than a grand one, for the city is not remarkable for its position, save that it stands to all intents and purposes on an island,

nor does it stretch towards heaven many of those stee-
ples, domes, and towers which are usually the ornaments
of the great cities of other nations. It looked to me
precisely what it was, a place of merchandise and accu-
mulated bricks and dollars, built for the need of immediately succeeding hours and for mercantile utility, rather
than for the pride and pomp of a great country desirous of
impressing the world with its magnificence. The ferry-
boats are managed to perfection, as to size, time, and convenience, and at an easy rate for freight and passage.
The Broadway is a fine thoroughfare, most dangerously
paved for horses, as is proved by their constantly falling
down in every sort of vehicle, light or heavy; but the
safety of the lives of men or animals is little heeded in
the United States, the dollar is the deity of the day,
and so long as that is fetched and carried, what lives are
lost, vehicles smashed, or knees broken, matters not to
the citizens of progression! In my way to the Clarendon Hotel I passed Grace Church, which was the hand-
somest place of worship of any that I saw in New York,
and the equestrian statue of Washington, of infinitely
better execution than any similar attempt to perpetuate
in stone the likeness of king or hero in the squares of the
metropolis of England. The seat and attitude of the
rider, and the action in which the horse is represented,
may well shame those English artists who have indulged
in all sorts of stony projections termed equestrian in the
mother country. This surprises me now more than it
did when I first saw it, and raises a suspicion in my
mind that a man of the old country, and not a citizen
of the new, must have sat a horse by way of model for
the artist's chisel, inasmuch as the seat of the statue of
Washington on his horse partakes more of the English

than of the American style (with the exception of some
of the officers of the army of the United States, among
whom I have seen grace and strength combined); the
New Country riders resemble in their seats tongs astride
a saddle, with their toes scarcely touching the sort of
fire-shovel or small coal-box things they call stirrups.

As I arrived at the Clarendon Hotel, I perceived that
the Ailanthus, as well as the weeping willow, grew by the
sides of the foot-pavements of many of the streets, plant-
ed in such situations for the purpose of protection from
the sun; and that almost all the lesser thoroughfares
leading from the Broadway and principal squares to
other parts of the town, were left in so impassable a state
of rough and neglected paving that the only conclusion
a stranger could come to was, that in this dollar-dealing
New World all the by-streets were the property of the
coachmakers, and, as a natural consequence, they were
kept in that dilapidated state to fracture the springs of
their customers' carriages and to increase a coachmaking
demand. In those streets also grass was growing up between the flags on the footways, giving to the vicinity,
state of carriage-way also considered, an idea of desolation that was disagreeable. Always being on the look-
out for birds, and there being one of the few pretty little
gardens in New York around a house opposite to the
Clarendon Hotel, the trees, shrubs, and streets, to my
zoologically-inclined mind, wore a solemn cast, for which
I could not account. On thinking of London, I discovered that, in my close contemplation of all things living and
dead, I missed the sparrows. In the streets and gardens
of New York, and indeed in the entire of the United
States, the bird called a sparrow in England does not
exist, and they have no house-haunting bird to supply its

place. In some of the gardens in New York may be seen an occasional fly-catcher.

From the hack carriage I was at last delivered to the Clarendon Hotel, and I found it an ample one, and for a wonder clean, and more free from that pigsty appearance of almost all the great hotels in the United States, caused by the beastly as well as to themselves unwholesome custom of spitting all over the floors, stoves, and fire-places in which this strange people so pertinaciously indulge. Oh! with what a comfortable sigh, after the confines of a ship, I arranged my dressing-case in a nice chamber, with its ample bath for hot and cold water within its own privacy arranged, and how I contemplated the comfortable bed—the very look of it was a narcotic;—but I had to prepare myself to dine at a table with many people, male and female, the *table d'hôte* being fixed for four o'clock, so I set about giving myself the appearance of an English gentleman, refreshed and ready for observation or adventure.

The hours for dinner in America are peculiarly inconvenient and erroneous in their arrangement, either for business, health, or pleasure; and the waste of time consequent on their fixture is so evident, that in a mercantile nation it is scarce possible to understand why there has not come about a wholesome reformation. It is curious, but in their hours of dinner and breakfast they seem to cling to the daily usages of a bygone era, as well as to those of what were originally, and which are now in England, the hurried meals of the working classes. In this they resembled or kept pace with their errors in the classification or names of birds and beasts, though of course among their very clever and well-informed gentlemen and professors there are those that equal our zoologists in England. Thus they generally and locally adhere to the first names given by the first Englishmen, Irishmen, or Scotsmen who landed on their shores, to the birds they first saw doing duty for birds of similar habits and size to those in the old countries from which they came.

The wood grouse in America is called " the phea-sant ; " the grouse of the plains " a chicken ; " and the bird a little larger than the English migrating quail " the quail," though that bird is to all intents and purposes the American partridge. All these birds I brought home in conjunction with the woodcock of America, which is not only of a different shade of plumage, but also less in size than ours. There is another bird of the grouse species which I did not meet with alive, larger than any of the rest, called the " sage hen," and every one of these I am perfectly satisfied could be acclimatised in this country. Indeed acclimatising they would not need ; the gold and silver pheasants show that it is not the peculiarity of clime nor gorgeous feather that makes the rule ; the food is the great difficulty. I have convinced myself that the food in England and the nature of the ground is precisely that which would suit the grouse and the partridge of the plains and woods of the United States.

To return to the method of life in America. The *table d'hôte* for breakfast usually commences at six in the morning and terminates at ten, and that for dinner from one in the day till five in the afternoon. Between whiles there is a most liberal tea and supper, so those who have the appetite and digestion of an ostrich can continue mastication all day long.

In many of the towns and cities the merchants dine at two o'clock, when, as many of their private houses are distant one or two miles from their stores or places of business, any one may well imagine the amount of time lost in going to and coming back from the points in question. Indeed, these gentlemen lose what in England would be called the best of the day.

Often did I attempt to combat these erroneous regulations thus existing among my hospitable friends, and jokingly point out to them not only their loss of time in locomotion, but the fact that they were fitter men for the affairs of merchandise or dollars when their cheeks were cool, their stomachs empty, and their brains free for calculation, than they were when oppressed by dinner, flushed, and jovially happy. It was all in vain! They would not permit innovation nor the introduction of reform even into their means of health or wealth; so I frequently dined when I was *not* hungry, and was made hopelessly to essay the digestion of the "devil and all his imps," who were dancing a polka for my punishment, for the sins of others, on my breast throughout the night.

There was no complaint to be made with the things mine excellently good and liberal host of the Clarendon Hotel caused to be set before you in the bill of fare for breakfast, dinner, tea, or supper. The choice was ample and the material good; but, hear it, O deity of the gastronomic realms, they harshly split and broil their woodcocks, and divest them of their trail! I remember explaining to one of these innocents in the culinary line, or rather fiends, according to the old saying that "Heaven sends food and Satan cooks," that the woodcock and English snipe should never be drawn, and that both should be underdone, or what the Americans call "rare." On explaining that the trail was the delicacy of woodcock, snipe, and red mullet, I thought the unsophisticated apple-roaster, for that they do very well, would never have recovered the end of his nose from its distortion of disgust, so shocked was he at the startling announcement of "the trail." I advise all travellers to New York to go to the Clarendon Hotel; it is clean and generous, attentive and good, and cannot be surpassed by any hotel, all things considered, that I ever had the pleasure of being in. The charges for room and eating are sufficiently moderate for the comfort enjoyed, and I was surprised at the price that was demanded, until I cast my eye over the cost of wine. That was infinitely dearer than in England, the wine no better, and oftener not so good, and of course that made very well up for the moderate suffering imposed for other things. The servants, male and female, in the hotel—indeed, it is the case throughout America—were mostly Irish; it is, and one not far-off day will terribly be, the curse of that splendid New World, that its own native hands are too democratically set up or foolishly and proudly insane to earn their honest livelihood by the fair and highly remunerative employment offered by their own country! What a lesson *that ought to be* to those English democrats and selfishly-prominent demagogues as to the boasted "blessings of universal suffrage," and what they call the "popular rights" of mankind. The old adage of the "beggar on horseback" is not inapplicable to the state of things a view of what ought to have been the labouring classes in America afforded me, and in my opinion the true interests of a country cannot be long maintained when all are masters and none are men; and a majority returned to Congress not by the wealth and intelligence of the nation, but by the houseless and ignorantly despotic, desperate and wandering class—a coercion exercised

over the free use of the conscience by numbers armed with
the revolver and rifle. To talk of the undue influence by
landlords in England over their tenants, their tenants
having many an interest *in common* with their landlords,
is as a mere drop of water in the sea compared to the
tyranny exercised by the "Boh-hoys" and "Rowdies"
over the freedom of universal suffrage; and when I read
the lucubrations of Messrs Bright and Cobden at the reform meetings, as they are called, when they hold up, I
fear, the disuniting States of America as a pattern to this
country, I am at a loss what to do—to marvel at the mistake of these orators, or to laugh at the impudence of
the attempt to lean a weight on the most fragile lath
that ever was put forth as a walking-stick for John Bull.
A man has but to travel in America and to mix with all
classes, to see the errors in the system of what may be
called the universal suffrage of an irresponsible people.
"A king can do no wrong;" that is a curious adage,
and may be a legal truth, though it cannot be a religious
one; and on closely regarding the law and its general
enforcement in the United States, the same may be ille-
gally said of the American people, "The people can do
no wrong;" and that fact will, eventually, I fear, be
"the knife" that will "sever the bundle of sticks,"
though accursed ought for ever to be the hand that severs
the unity of a splendid citizenship that can only fall to
pieces by suicide, or by an insane intoxication produced
by too much liberty. Liberty is a beautiful thing in the
social as well as the political or public system throughout the world; but if a man takes too great liberties even
in private society, liberty destroys the delicacy of life, and
leads, if unchecked, to demoralisation. The same result
is, at this moment, hanging by a thread over the pros-

perity of America; the people are drunk with liberty
already, to the destruction of the best of all blessings—
the ample and remunerative employment of the indus-
trial poor.   In the States two interests have clashed, and
are clashing, fomented, I deeply regret to say, by clever
and popular writers from England, one of whom was
made to believe, during his visit to America, that slave-
owners marked their slaves in the United States by
" knocking out a front tooth; " and in the other an authoress assured us that slave-owners might always be
recognised by the size of their right hand, the muscular
proportions and weight of which " were increased from
their constant habit of knocking down their negroes."
The slave question, which has already led to so much
bloodshed, as well as to the pilot balloon sent forth to
test how the wind blew, in the shape of " Old Brown,"
is the *béte noir* on the horizon of the United States. With
that looming on them at this very moment, and the intoxicated state, in regard to liberty, of the masses,—the
numerical insufficiency of the standing army, though
splendidly officered by soldiers and gentlemen, sons of
America,—the fact that that army is a strange or foreign
force, comprised of foreigners and not of American citizens,—and the contempt in which many of the laws of
Congress are held, punishment uncertain, and immu-
nity even for murder purchasable by money,—any man
with a head on his shoulders who has travelled in America cannot but fear that the mere weight of a feather
thrown into the scale may bring about a collision that
will lead to the most terrible results.   Many friends of
mine in America, among them a widely experienced and
most gallant and able soldier, assure me that they do
not dread disunion, for all the confusion at hustings,

murder, and even petty civil wars that have been, and are still, notorious. According to their opinion the men of the slave and anti-slavery States, when the hour of disunion arrives, know well that in disunion lies an individual ruin, and they will do anything rather than cut the withe that binds them unitedly together. In this opinion, may Heaven grant that my friends in America may be correct; but, as my sporting narrative proceeds, I shall have it in my power to show many a little circumstance, slight perhaps in itself, that tends, when considered in the aggregate, to show grounds for apprehension that things within America are far from peaceful in their attitude, and that there is nothing so dangerous as liberty insufficiently restrained, or a suffrage so universal that property and life lie at the mercy of an irresponsible multitude.

CHAPTER III.

DESIROUS of seeing as much as I could of New York, and the life and customs it contained, and anxious to be introduced to all those resident gentlemen to whom I had letters, as well as to make inquiries with regard to my journey, and the outfit needful to the plains, I remained a week at the Clarendon Hotel. In the way of amusement, and during their short stay, Sir John Rennie, Mr Brown, and myself were in the habit of driving out after our unconscionably early dinner to see all that could be seen of roads and buildings, for, after dinner, as I before said, there was an end of business. As far as the city went, excepting its size and bustle, there was very little to be seen, and nothing to marvel at but the monstrous size of the hotels, at one of which there was an apparatus for craning the occupants of the higher buildings up to their apartments, and in this hotel, as the story goes, "a gentleman not feeling well had gone to bed, and ordered a cup of tea to be brought to him. This order entailing considerable travel or a sort of treadmill on the (*Anglicé*) waiter, that gentleman, being of course free in every sense of the word, hated trouble, so he put the order on one side, and the tea was left unapplied for. Six weeks after the written request the card for tea turned up, and the general or colonel behind the bar having issued a desire that it should be furnished to the gentleman in Room 742, a waiter took it up, and found only the remains of a customer; for the gentleman who once wished for tea could not drink it, having no stomach for it, as he had been dead six weeks."

I went to see this huge building, and found that it was not only an hotel, but also a sort of inclosure for the society it temporarily contained, insomuch as a promenade of its inhabitants took place every evening in the long passage round the hotel; a very interesting proceeding, I should conceive, to any occupant desiring quiet. Many of the doors of the respective apartments opening into this public highway, of course the measured tread of people arm in arm was, to the imaginative listener, as the march of soldiers, and while it continued rest was out of the question. The larger the hotels, the greater number of loungers there are in the entrance-halls, seated in every attitude not designed by nature to give them rest — their shoulders, and at times the backs of their heads, being those portions of the human frame applied to chairs and benches used by Englishmen to be sat on in a more natural way. In one of these large halls, where everybody is lounging, smoking, and chewing tobacco, if all chance to cease speaking together, in that momentary silence there is precisely the same noise, but with a more filthy effect, that there is beneath a large rookery at night. The dirt from the lips of men, and the same from the rooks, in noise, quality, and quantity, are very similar.

The Broadway is certainly a very fine thoroughfare—that is, when compared to the deserted by-ways and grass-grown streets which lead from it; but in London it would have nothing but its length to make it remarkable, for it is not to be compared to Piccadilly. At one end of it, like Piccadilly, getting towards the City, it is confused and crowded by numbers of drays and other descriptions of carts, not forgetting those frightful nuisances the omnibuses; and even throughout its length these huge transporters of an irresponsible people cumber the highway, and bully any other vehicle less than themselves. The street rails which are laid down for one particular class of these cumbrous nuisances are the greatest eyesore and hindrance to other travel that can possibly exist. Thank Heaven, London is not yet grown so tyrannically democratic as to tolerate such an incubus on the locomotive freedom of the world. How intensely I have been amused at the things the streets afforded, so utterly at variance with the nonsense talked about the " national and beneficial equality of the American people ! " Lumbering along rails, tyrannizing over all other wheeled conveyances, there you have a cheap omnibus; and when I demur at being shut up with " Rowdies " and " Bohhoys," all of them spitting in chorus between their knees, my democratic friend, who shows me the town, assures me that in the land of freedom *there can be* but one

price and *one* vehicle for *all.* "Well then," I remark to
him, "wonder not at my surprise, when, fully impressed
with this levelling idea on coming into New York, I saw
drawn up at the side of a street or square what I deemed
to be a lot of little Lord Mayor's coaches, bedizened with
plated metal, silver, lace, and other profuse decorations,
and learned that these were two-horse 'hack carriages,'
on the stand *for public hire,* and not for the locomotion of
civic dignitaries, and above all, that the cost of the hire
of these *public conveyances* put them utterly and completely
beyond the reach of any but the better and richer classes."
Curious this; but I shall in other places have to touch on
the boasted and said-to-be beneficial and national equality again.

I must, in passing, recount one or two of the many instances of the pleasantries of the omnibus, and the delights for the well-educated and moral traveller contained
within their tobacco-juicy panels. The day was intensely
hot, and humanity in the streets, as wasps, unpleasantly
busy. The omnibus had become crammed to suffocation
with fellows who perpetually startled you with the idea
that they had spat on you—a fate which is to an Englishman as filthily insulting as it seems to be soothingly
pleasing to an American. They did *not* spit on you,
however; they missed your knee or the flap of your coat
to a hair's breadth, and only put patterns on the floor between your feet. Had the Americans been half the shots
with firearms they proved to be with the juice of tobacco,
I should have gone home a defeated sportsman; as it
was, they spit with infinitely better aim than they shot,
though generally they were very good with the rifle, and
fair with the shot-gun, but by no means superior to
other nations. Well, then, the crowd in this omnibus

was excessive; every seat was occupied, and the passage
from the door behind to the spot immediately beneath
the driver was choke-full of people, in the impossible position of the winged heads poetically designated as che-
rubims—they *could not* sit down, for there was not the
wherewithal for the restful accommodation. Men stood
on the steps behind, and others clung to their skirts, and
the Rowdies and Boh-hoys seemed to be the chief of the
Jehu's congregation. Immediately beneath the driver,
in the remote corner of the carriage, there sat an excessively nice-looking old gentleman (offering a wide contrast to those called his equals), who, being unable to
resist the state of things any longer, knocked at the top
of the omnibus by the hole through which the driver receives payment, and called the lord of the whip's attention. Having been answered by a corresponding vibration
to prove that the desired attention was obtained, " Driver," said the nice, respectable old gentleman, " driver,
your carriage is fearfully crowded; half New York seem
to have stood on my feet for the last twenty minutes; I
have neither room nor air. How many more do you
mean to admit?" In vain did the meek grey eyes of this
nice old man watch the hole above, through which the
money passed, for a consolatory reply. No sort of answer for a length of time *was* given; at last a thundering
noise of knock-me-down-knuckles rattled on the roof
above, and a bearded chin (there was no room for more
of the face to which it belonged) thrust itself with considerable force into the hole, and, wagging its hirsute goat-
icity like the tail of a dog when shaking off the water,
a concentrated but nasal voice replied, " How many
more *am* I *going* to let in? Just as many as I please!"
The last of these words were widely divided, and they

were each given a concentricity which it would be impossible to imitate. The nice old gentleman collapsed
at once, and cast his eyes round at every decently-dressed
person in search of commiseration, but every soul burst
out in laughter, and the driver seemed to be hailed as the
true type of American independence.

While on this topic, I must also relate a most amusing
anecdote of the Broadway in New York, which befell a
young gentleman gathering *useful notes* for an historical
publication during his stay in the city. He had heard of the
immense number of " Boh-hoys " who were supposed to
be not only in New York, but extensively and numerously
scattered all over the United States, and having been
very truly but slyly informed that they were, in times
of political excitement, a very influential body, he very
naturally wished to learn something of their social history.
For the better information of my readers, it may be
necessary to state that an American Boh-hoy is about equal
to one of the *lowest* English cabmen. To this end, and in
search of agreeable knowledge, the young historian
walked into a large " store," its contents being utterly
disguised by the innumerable lettered annunciations supposed to explain them on its front, and saw at the further
end of the shop the back of a chair and a portion of the
figure of a tall man, as from the waist to the feet, standing, as it appeared to him, after the manner of a fly upon
the ceiling. There was no head nor shoulders to be seen
on account of the chair's back, but the portion of the
figure before alluded to was, as it seemed, appended from
above. There was a dead silence in the shop, rather
oppressive to the young hunter after knowledge, when,
though he trod on purpose loudly and coughed repeatedly,
the upside-down portions of the human figure, arising

from behind the back of the chair, never moved. On
went the youth, for he thought it too late to turn back
until he rounded the side of the chair, and, to his astonishment, beheld a tall, gaunt, three-corner figured man, above
six feet five inches in length, seated on his shoulders. I
would here make a passing observation that Americans
can, and with apparent comfort to themselves, not only
sit on their shoulders, but I have seen a citizen, for ease,
coolness, and comfort, sit on the back of his own head on
a desk-stool, all the rest of his person tending upwards
towards the realms above. The lips in the head (that, as
he rounded the side of the chair, turned with an angry
sort of stare towards him) were bulged out with a quid
of tobacco the size of a duck's egg, and were actively
employed in smoking a toothpick. The stare cast on
the intruder seemed to say, " Wall, I guess if you want
anything, you 'll speak some ? " So the intruder, thus
mentally apostrophised, blurted out a " Beg pardon;
perhaps you 'll be good enough to afford me some intelligence ? " " Hum, hum ! " replied the toothpicked head;
" wall, I might try." " Thanks then, sir; are you acquainted with the society of ' Boh-hoys ? ' " The sharp,
dollar-seeking eyes of the tall man twinkled rather slyly
at this, and he replied, " Guess we all know 'em." " Indeed, sir," rejoined the querist, " then you are just the
gentleman I want. The Boh-hoys are a numerous body,
are they not ? " " Guess they are." " Could you introduce me to one ? " The legs drew themselves very
slowly from their lofty position, and with much the action
of a snake, coolly and slowly the body and head rose, as
the waist and heels came down, when, sitting bolt upright, and eyeing the intruder slyly, the tall man said,
" It don't want much introduction." " Can you, then,

show me one of this society in the street; there are numerous people of all sorts passing your door ?" " Yes," replied the tall man, still eyeing his young friend very slyly, " guess I can;" when he took his toothpick from his mouth, and slowly advanced to his shop-window. He stood there for a few moments taking a tall view of passing people, while the little would-be-historian regard-ed the inhabitants of a free world from a much less lofty position, keeping his eyes to the better-dressed pedestrians on the pavement, while those of his companion, perhaps, looked beyond the gloss of fashion and sought out its equals, but in rags and tatters. The youth was suddenly startled from his more lowly place of observation, by the gaunt hand above him emphatically striking the butt of his toothpick against the glass, with the observation, " That's a Boh-hoy." " Hey ! what ? who ?" exclaimed the delighted historian ; " the gentleman standing to speak to another just by the door; the smart man in the high hat ?" " No," cried the tall store-keeper, at the same time turning from the window ; " the Boh-hoy with his legs a-straddle, in his shirt-sleeves, upon the dray." " May I speak to him ?" ejaculated the little man, in some surprise. " Guess *you may*," cried the tall one, slowly going to his chair at the further end of the shop, and the historian's legs carried him rather convul-sively into the street. There was a general stoppage of all vehicles at the moment, so the literary youth went across the Broadway towards the member of the influ-ential society ; but not feeling quite sure of a polite re-joinder to any close inquiry as to personal affairs, he addressed the Boh-hoy, as an approach to further con-versation, thus, " Sir, I want to go to Seventh-street." " Then why the devil don't you go there ?" was the startling rejoinder. The wheeled detention in the thronged thoroughfare then became freed once more, the Boh-hoy and his dray moved on, and the young historian had to skip for his life, in order to avoid being crushed by vehicles of every description. The drivers of every omnibus, dray, cart, buggy, hackney carriage, or gentleman's carriage, sit on their driving-seats, or stand on their drays, with their legs "straddled" as widely as possible, and their arms similarly spread, a rein grasped firmly in each fist, and the mouths of their horses the chief fulcrum by which the load behind them was propelled.

At New York, in driving out after dinner, we went to look at the public park that is being made, and in doing so we drove some little way on one of the most dusty roads I ever saw, greatly frequented at that time of the afternoon by gentlemen amusing themselves in those vulgar-looking gigs that are made very light, and with high wheels, to be pulled along by their fast-trotting horses. I scarcely know which of the two things to dislike or which to laugh at the most, the shape of their vulgar-looking vehicles, or (to an Englishman) the ugly and ungraceful position of the drivers, who sit the very reverse of our English position, all abroad like a spread eagle, with legs and arms as wide apart as possible, and a rein clutched in each fist, hanging on to the bit in the distorted jaws of the mouthless horse, and making the reins and the horse's head do the duty of tug, trace, and collar. In short, the harness is only needed to hold up the shafts, and to keep the animal from escaping, all stress being on the rein, and from the arms of the irresponsible driver, who holds on like grim Death for his own peculiar pleasure and the probable destruction of mankind, and can only stop by letting go the reins.

The pace that these "trotters" drive at is awful, when, as their horse has no mouth, and is absolutely by his jaws testing the utmost of the weight and strength behind him, if they come suddenly on a quiet gentleman taking a digestive drive in the dusk, the probability is that he suddenly finds himself let down from having a wheel cut off, and before he is quite sure as to the nature of the shock and appending catastrophe that has befallen him, from the midst of a cloud of dust above he hears a nasal, and, were it not for disaster, an amusingly insufficient apology of " S'cuse me," and the trotter flies away with undiminished, or rather an increased speed, lest he should be called on to pay dollars for a death or destructive difficulty.　　That accidents are not more frequent I only wonder.　I saw but one the whole time I was in America, and that was at St Louis, when a trotter caused a difficulty by overtaking a baker's cart.　When Mr Campbell, with whom I was driving, and myself reached the spot, there was a sort of pile of rubbish, beneath which there seemed to my astonishment something alive, and at last the baker scratched himself to light, after the semblance of a mole, in great doubt as to what had befallen him—whether he had been smitten by a flash of lightning, or blown up by one of his adulterated loaves.　　His personal injury amounted to no more than a scratched face.　The astonishment of Americans when I told them that, with all my predilection for sport, I cared nothing for a trotting horse, or matches against time, was great.　They seemed to think it was the most interesting of all equestrian perfections.

During my stay I visited Fulton Market, and found it well supplied with meat, fruit, and vegetables, and with game when the weather was cool.　The beef was good, but not so fine as in England ; vegetables and fruit excellent ; the mutton very middling, and the fowls nothing like so fat or large as those in the Old World. As to the venison, from the common deer of the country, it was killed without care as to sex, age, or condition, and I saw none that in England we should call fat, and none ever came to table with what fat there had been on them, for in cooking it was all roasted off. In addition to this, they kill the bucks too late, or when "the rut" is coming on, which of course militates against both fat and flavour. The meat of the deer is fine in grain and good to the taste, and, if fat, I have no doubt but that the venison would be excellent. As to fish, the market (if I may judge from what I saw at the hotel as well as on private tables) was badly supplied; the only good sea-fish I ate there, and that was but once, was called the "Spanish mackerel." It seemed in flavour, size, and hue to be much the same as those caught on the coasts of England. The best fresh-water fish that I tasted was "the bass," which is, in fact, the American perch, and I believe grows to 4lb. and 5lb. weight, and occasionally more. They have a less fish, that they call the "black perch," which is by no means a delicacy; and also a fish they call the "crappeè," shaped something like the English bream, which has very little to recommend it. The catfish, which some of the Americans are very fond of, when dressed resembles a bad eel. Excellent sport can be had in the spring and fall of the year, by spinning a minnow for the bass and crappeè fish in the tributaries or back-waters of the Missouri and Mississippi, and the black bass will rise freely at a fly. In the lakes of America there are the trout and what they call the lake salmon, and the common pike and other fish ; so that the

angler, though in the parts I visited there are no salmon
rivers, need not despair of plenty of amusement. The
catfish, weighing 60lb. or more, can be caught with a
night-line, and at times by the spear. There is also a
fresh-water dogfish, who shows excellent sport in his run
at the minnow, but he is useless for the table.

I endeavoured to bespeak of Mr Robertson, the chief
poulterer in Fulton Market, some live feathered game to
take home to England, of the three sorts, the wood-grouse,
the prairie-grouse, and the quail or partridge; but the
obstacle to his receiving them alive was beautifully illus-
trative of the idle habits of his countrymen. He told me
that perhaps they might be caught alive—the quails certainly—but the difficulty was, or, in short, the impossibility was, in finding *any American*, though paid to do so,
who could be trusted to trouble himself sufficiently to feed
the birds in their transit from where they were captured
to New York. They were sure to be neglected on the
way and starved to death, unless in charge of a slave or
a German servant. So convinced am I, however, that
these birds, as well as the Deer, would thrive well in England, that I shall continue my attempts to get them across
the Atlantic.

A very few days gave me a pretty good insight into
most of the manners and customs of Americans; and
among some of the things that struck me as extremely
odd was the fashion among the ladies of wearing their
hoops and crinolines! I saw New York for the first time,
be it observed, at a period (September) when the ladies
of fashion or in the best society were out of town. When
first I walked into the Broadway at the fashionable hour
I had a strange sensation as if my head was turned, and
had I had an appendage, like the sailor in the song, "I

should have chewed my pigtail till I died;" or as if, through some anatomical freak of nature, women's waists, bust, and head, in the United States, had been set on the wrong way, and their "bustles," instead of their "stomachers," been on before; or that they looked over their heels instead of over their toes to see if an admirer approached them at unawares. In a short time I recovered from my perplexity on finding that the mistake originated from the fulness of the dress being made to stick out to the front of the figure instead of behind, and that it was this strange method of personal decoration that induced the supposition that women in the United States were perpetually "hind before." Some of the ladies I saw were very pretty, and there were some dangerous "twin invaders of domestic peace" in the shape of feet and legs (I have a strong suspicion that female America is famous for them) that made me mentally draw comparisons. But, England, dear old England! be not jealous; for, by all that a soldier and gentleman holds dear in leg and foot, you were not outdone, and in teeth, speaking collectively or nationally in that particular, you stand unrivalled.

In this passing glance at the social arrangements of the United States there is a fact which I approach with sentiments of the greatest delicacy, because it treats of a state of society which, while it is very beautiful, is nevertheless one that opens a wide door to abuse as well as to a state of Lynch law, from which, however useful at times, civilisation in any country should be exempt. I allude to the great liberty given to girls as to their unchaperoned communication and association with men. They are permitted to attend balls without a *chaperon*, and to ride, drive, and to walk without them—a state of things immensely comfortable, and to the recipients of such freedom delightful; nevertheless, while I admit the happiness of such regulations or irregulations, I must not shut my eyes to the Lynch law or serious difficulty that very naturally, on such a state of freedom, as freely and according to nature, arises. The Lynch law to which I allude is, that if, when two persons are thus beautifully associated, the man commits, or is even supposed to have committed, an error, the relatives of the girl take the law into their own hands, and, without a desire to conceal the scandal and shield the girl, they at once give to the matter a terribly public complexion, by meeting the lover in the streets, and shooting him down as if he were a mad dog. I admire immensely the faith that mothers have in their daughters—I love the free intercourse of the spirit thus permitted—but I wonder not at the natural consequences that unlimited intercourse, or a notion of injured honour, or even mere jealousy, may occasion.

During the time that I was making these observations I had been delivering my letters of introduction; a great many gentlemen were out of town, but those who were at home received me with the utmost friendliness and kind attention. If I had seen the lower classes rudely intoxicated with liberty, and many of the Boh-hoys disgustingly obscene, still had I put them by the side of the inhabitants of the London Billingsgate in blackguardism there would have been but little difference. A Billingsgate man by the side of a Boh-hoy, however, would have had this great advantage. In the former, though there might have been all and everything to condemn, there is no hail-fellow-well-met assumption as to being on a footing with his betters; while with the last there is that revolting assumption, and the thief who had stolen a crust of bread from a child's mouth, or her mite from the poorest widow, and who would meanly cheat to the fraction of a "dime," would, in his unwashed depravity, have held out a hand to a gentleman as if he had been the equal of chivalry and the personification of an honest man. The real American *gentlemen* cannot be surpassed in high-toned feeling, graceful sentiment, urbanity, or courage ; and though it is the fashion among some of my country-men to deny an aristocracy to America, I know that there is as high a feeling among her born gentlemen, and in the upper classes, as there is in any class in England, and as much difference between the American peer with-out a title and the American "villain" or Boh-hoy, as there is between the earls of England and the lowest of the rabble ; and every hour this fact will become more manifest. The position taken by the class that may justly be termed "the Aristocracy of the New World" is, to my mind, at this moment peculiarly dignified and graceful, while at the same time I cannot but think it bodes no good to the national prosperity, and proves the errors of universal suffrage. The upper classes stand aloof from the political world ; they feel that they are unrepre-sented in Congress, and that they at present possess no power to compete with the overwhelming will of the masses ; that numbers, and the hasty naturalization of foreigners, and a fluctuating, wild spirit for speculation, have usurped the functions of all who are possessed of real property ; and that until this very false state of affairs rights itself, religion and law, education, wealth, and strength, must be at a discount, or stand trembling on the eve of destruction.

I will not attempt to enumerate those from whom I received the utmost kindness and attention, lest I should unintentionally omit one, and thus appear a bankrupt in gratitude. Some, who came prominently into my plans, I must of course individually refer to as my narrative proceeds, and I trust with justice; it is at present enough to assure my readers that I have a deep debt of obligation to all my transatlantic friends.

Having got over the impression at first made on me, that the tradesmen of New York were the most brusque and ill-conditioned set of fellows I ever met with, who seemed utterly to have mistaken the position of tradesman and customer, and to regard you, when you came to make purchases, as an intruder who was seeking a favour at *their* hands instead of doing one to them, I went to Mr Duncan's store in the Broadway to order what was henceforth to be called "the prairie chest," and to get some wine and brandy. I found him politeness itself, and, from the commencement, a strong contrast to many of his compeers, who so foolishly and erroneously try to impress you with their equality in state and station, by disfiguring themselves with a rude manner which really does not pertain to their nature, and which melts away on better acquaintance. From Mr Duncan I received much good counsel in the fitting up the chest. It was obvious to me (though no one advised it) that to travel a month or six weeks in a desert it would be well to carry with you some choice provisions to fall back on if the weather, sickness, or any incidents prone to happen on plains, the abode of malaria, as well as of a savage and treacherous people, should militate against supplying the camp with food by hunting, and that these provisions should be packed in as small a compass as possible. I selected hermetically-sealed cans of ready-dressed beef and gravy, and beef-gravy and vegetables, and the same in regard to mutton; but each of these things, when

eaten cold or hot—the heat was at once procured by
plunging the can into boiling water—proved to be made
with too much richness for the peculiarity of the fickle
climate, and liver-attacking nature of the fever and ague.
The cans that stood me in good stead were those of potted
and ready-dressed chicken, as well as potatoes in a can
by themselves, cherries and apples ; to these latter, the
fruit, I mainly attributed my resistance of the fever, and
were I to seek Mr Duncan's store for a second collection
of the "prairie-chest," I should, in the place of the beef
and mutton, have more chicken, and some cans of preserved salmon. Whatever wine the traveller has been
accustomed to drink in other countries, and that will bear
carriage and yet be fit for immediate use, that also he
should take with him. I hardly ever drink anything but
sherry, and Mr Duncan supplied me with some very
good, though not in sufficient quantity, and I had to
increase my stock of sherry at St Louis, but of a quality
not so good as his. I also took with me a couple of
bottles of port from Mr Duncan's store, in case of sickness requiring such a restorative. For a man's individual
consumption on the plains, during a month or six weeks,
the quantity should be three dozen of sherry. Of course,
a certain amount of good brandy should be also in store ;
though with the liver affected, as the plains affect it, I
am convinced that brandy, or any spirituous liquor, and
more particularly tobacco, are the worst things that can
be taken. Unacclimatised as I was, I had less fever than
my men, and no ague; and from the enervated state of
the Americans generally (they are most indifferent walkers
and very easily tired), I am perfectly convinced that they
undermine their manhood by the unlimited and filthy
use of tobacco in smoking and chewing, and that the

copious expectoration in which they seem delightedly to
revel, to the misery and disgust of all their fellow-crea-
tures who happen to be near them, is detrimental to their
digestive organs, through the waste of the gastric juices,
and that the citizens, if they took a turn for soldiering in
foreign countries, would never live through a long cam-
paign.   Several of my friends in England advised me
(though they knew that I never smoked a cigar) to be
sure to take to the desert a good store of tobacco, and
they smiled at the idea of my being able to do without it.
I never touched either cigar or pipe, and when I saw the
condition to which the inveterate smokers and chewers
of tobacco in the United States were reduced, when exposed to the terribly uncertain climate on the plains, or
under violent exercise, I never could have been better
pleased than I was with the abstinence from tobacco of
any kind in which I had ever persisted.

It will, I am sure, amuse my readers in both countries
when I give them a few of the cautions received by me
from my supposed-to-be-" experienced friends " in England, on the eve of my departure from the old country.
For brevity's sake I make out the following list of kindly-
intended but not very practicable advice.

" When you get to America, all you will have to do is
to attach yourself to some trappers going to the desert,
do as they do, take no tent, no waggon, no knife and
fork, no attempt at any better fare than they have, for
they will despise you if you do, and *while you are buffalo
hunting*, they will, with their hatchets, build you a comfortable hut *with leaves and the boughs of trees*, and then
you 'll all dine and smoke together round a fire !

" You will find buffalo a few days' journey beyond New
York, and elk and deer in any quantities ; and if you can

meet with any tribes of Indians, or, better still, those excellent fellows the Pawnees (Murray, in his book, says he remained months associated on the plains with this tribe, and never passed so delightful a time in his life), you may safely rely on their fidelity and honesty, and they will take you with them and show you all the fun.

"It is no use your taking English dogs; the plains are so thick with thorns and bushes, heavy grass and jungle, that an English dog could not go a yard.

"You'll suffer terribly from mosquitoes and a fly with a bayonet-shaped proboscis, and nothing you can do will keep them off. The mosquito-curtain, of course, you will have, but the mosquitoes get under that.

"Take no bed. Take a buffalo robe and your blanket; that is all you require, for the Indian summer is the most beautiful thing in the world—very mild and yet bracing, the heat of the sun tempered by a mist in the sky.

"Take some brandy—a little; you will find plenty of whiskey in the country, and that with water is all that you will require. In short, your buffalo robe and blanket; rifle, shot-gun, and tobacco-pipe, with an association with Indians or trappers, will carry you in safety wherever you wish to go; and if on starting from England you put £150 in your pocket, or have credit in the United States to that amount, it will be all that you can require.

"By all you hold dear, and for the safety of your life, avoid the river steamers. The captains are hard-drinking, dangerous, go-a-head fellows, who, with a stomach full of ardent spirit, exclaim on starting, that, reckless of the safety of their vessel, the merchandise in it, or the lives of the men, women, and children entrusted to their care, 'they'll go ahead if they bust their boiler;' so for

4

Heaven's sake keep clear of them. On board their ships all is riot and confusion, drinking and swearing; you'll be made to sit up smoking all night, and ten to one but some great bearded, disgusting fellow gets into your berth along with you, with the exclamation of ' Stranger, I reckon this here crib o' yourn 'll hold us both,' and that will be agreeable! As to the comfort of dinner or any meal on board the river steamers, if you do not rush in among the foremost and quickly seize whatever dish is nearest you, the hands of all your neighbours will be tearing everything to pieces, and, in half the time it takes to tell it, nothing will be left on the table. So look out."

Advice, and such advice as this, was given to me; the amount of worth which that advice contained my readers, by my succeeding narrations, will learn. Faults enough in America there are, as there are in any other country, but over those faults there are a thousand virtues; and if I give not both their due I do not deserve the name of an English gentleman. I write not to pander to a morbid taste in England, which delightfully revels in the abuse and condemnation of others, and I attempt not to make a tale wondrous or popularly palatable by false assertion. Truth is not only more beautiful, but it is more astonishing, than fiction. I wish many of my American friends would think it so; and the faults of the splendid New World, as well as the perfections of it, shall here be most faithfully recounted.

CHAPTER IV.

AMONG the gentlemen in New York who so kindly rendered me all the information in their power was Capt.
Marcey, the author, I believe, of that very useful little
book entitled "The Prairie Traveller," and published
by Harper Brothers. If my memory serves me correctly, it was this gallant friend of mine and successful hunter who suggested my going from New York
to St Louis, thence by the Northern Missouri Railroad
to Hudson; from Hudson by the Hannibal and St
Joseph Railroad to St Joseph; from St Joseph by river
steamer to Omaha; and from Omaha (192 miles) to Fort
Kearney, where there is a station for the troops of the
United States army. In this route there would have
been settlements to within thirty miles of the fort, and

on the road, and only nine miles from Fort Kearney, on Wood River, very good wildfowl shooting, and buffalo adjacent to the fort itself. I was further advised to buy waggons, harness, and saddlery at St Louis, and send all my baggage with the waggons by steamboat to Omaha, steamers plying every day between that place and St Louis. After much consideration on several points offered to my notice, I resolved to start for St Louis direct by rail, and to take for myself, my servant, and my dogs, tickets through, a power being given to the takers of such tickets to break the journey at any place on the rail they pleased, and to consume as much time on the road as might be convenient. It seemed to me necessary to the comfort and condition of my dogs that they should have time for rest, exercise, and food, without being shut up in such accommodation as the train afforded for from four to five days. On making further inquiries, however, I found that I could only take tickets for myself and my servant, dogs not being considered by the sagacious heads of the railway companies as animals of freight for which any accommodation should be given; and that *if* I took my dogs it could only be by making a private bargain, and on such terms as the Boh-hoys or baggage-masters might please extortionately to demand. In one or two instances only (and I must think accidentally) I found these servants civil; all the rest were as bad specimens of the genus Boh-hoy as the back slums of the larger cities could afford, and bent on every species of extortion and theft.

On finding myself placed in this dilemma, I went immediately to Mr Hoey, one of the managers of Adams' Express Company, and applied for his assistance. He very kindly and with the greatest good-will gave me

a letter, to be shown to all the express messengers on
the line; but, with one exception, they paid no more
attention to Mr Hoey's desires than they would have
done to those of any one of whom they knew nothing.
And here again I learned that every American deemed
attention or obedience to his superior, whether in education or office, as degrading to the state of his free citizenship, and beneath the consideration of men of a nation
possessed of universal suffrage.

Having made the best arrangements in my power, I
took tickets for the journey throughout, with power to
stop where I pleased, and met my dogs and George
Bromfield on Wednesday, the 8th of September, at the
station in New Jersey, a little before 4 p.m. Here, then,
my troubles on these miserable railways commenced;
the insolent baggage-master, although I showed him Mr
Hoey's letter, said he could not take my dogs, for he had
no place to put them; to which I replied, "Then I will
not go." After some demur, and the train on the eve of
departure, he pointed to a little sort of cabin on the train
(there is no such thing on an English line), and said
some of them must go there, and the rest with my man
in the baggage-van; and to my horror I saw Druid dragged in by a fellow in rags whom he had never seen before, who was told to hold him and one of the other
dogs; and while the carriages were in motion, somehow
and somewhere or other, myself and Bromfield and the
other dogs got in, and were soon progressing in a most
unsatisfactory way. This train was to convey me only
as far as Philadelphia, and to get there I had to take the
boat across the river, and then another train on all night,
alleged to reach Altoona at 6 a.m., but we were hours behind time. My object in reaching Altoona at that hour

was, that I might have daylight for the transit and splendid views over the Alleghany Mountains and through the forests, and an opportunity of observing a curve in the line of rail that, not many years ago, or even now in England, would have been regarded as impossible; dangerous it must be, as safety depends entirely on very reduced speed.

On arriving at Philadelphia tickets were delivered to us on being severed from our luggage, which would enable us to claim it on reaching the Philadelphia Hotel; and I then proceeded to walk to the hotel, because the streets at night were not only crowded but particularly filled with that influential class the Boh-hoys, and I desired to be careful of my dogs. I then proceeded through the streets in front of my four-footed companions, George attached to their chains in the rear to keep them steady. All at once, from behind a dark corner of an adjoining dirty little street, and on seeing my dogs at my heels, there rushed about fifteen or twenty of the lowest Rowdies and Boh-hoys, one of whom—I suppose the leader —stopped short, and, looking me full in the face, cried out in a most insolent nasal twang,

"Hum—hum—hello—stranger—are you a dog-catcher?" At this question his companions roared with laughter, and I replied,

"Yes, I am; I catch all the nice and civil dogs I can find, give them lots of grub, and treat them well; but *if* I come across an insolent puppy like yourself, I just about lap into him uncommon; so you'd better cut!"

The manner in which I gave vent to this rejoinder so fascinated the spokesman Boh-hoy's followers that they in turn roared with mirth, in *my* favour, and I passed on without further molestation.

Arrived at the hotel and station at about half-past nine at night, I conveyed my dogs to a safe corner on the platform at Philadelphia, and ordered George to hold them short by their chains, that Druid or Brutus might not bite some of the citizens, and render their gait, at all events, less free. I had not been arrived long before it became known who I was, and I was at once accosted by a railway official, I believe, who astounded my peacefully-conceived ideas of his nation by then and there, in due form, introducing me to seven generals, five colonels, and three majors, all accidentally got together, not one of whom looked as if he had heard the report of an enemy's cannon, seen the glitter of a drawn sword, or had even the knowledge of how to wear one without its getting between his legs and throwing him down. As to drill, it seemed to me as if they had never in all their lives or previous to having attained their high military rank been told to hold their heads up or march across parade. They were all very kind and civil, however, and after having welcomed my arrival with the utmost good-nature, under great apparent curiosity they proceeded to cluster round, inspect my dogs, and ask questions of George Bromfield.

"Beg your pardon, gentlemen," I heard him say, "one or two of my dogs don't like strangers; have the goodness, gentlemen, not to come too near, nor touch them."

"Oh," exclaimed one of the bystanders, no doubt high in military rank, and with a good deal of self-important assumption, and dressed in a very smart waistcoat spangled with stars, "no dog ever bites me; I can be friends with them and make them know me at once." So saying, he stepped out, and reaching his hand toward Brutus's black head, he was about to pat him, when, knocking

his arm on one side, the retriever flew up and seized him
fast by the smart waistcoat, the greater portion of the
constellations on which no longer remained fixed, but
commenced to be revolving stars, at length going out by
shooting towards the earth. It seemed that all this ex-
perimentally-inclined gentleman's friends rejoiced in the
scene, and in the midst of a roar of good-humoured
laughter he vanished, in possession of only the back of a
waistcoat, amidst the crowd.

Time for a fresh start having arrived, and having paid
the insolent baggage-master ten dollars for the transit of
my dogs so short a distance, and without which he had insolently refused to take them, I helped them into another
robbery car, and, leaving George to travel with them, I
seated myself in one of those detestable long carriages
on a short seat at the end, with my face at a window.
I had not been in that seat long when I heard moans of
pain immediately behind me from a man seated by the
side of a woman. Later in the night the man asked me
to exchange seats with him, and to let him and his wife
sit on my seat while I took theirs, for that he had lately
lost his leg considerably above the knee, and was not
only in great pain, but so unaccustomed to move on one
leg, that unless he was quite close to the door, he feared
he should not reach it in time to get out at his place of
destination. On hearing this, I directly acquiesced, only
stipulating that if I gave up my seat to them they would
assure his wife's seat by the other window to me, the
train was so filthily hot, and disagreeable from the beastly
ejections of tobacco. To this they thankfully consented.
Seeing that the man was not only heavy, but much exhausted, I then got up and assisted his wife to move him
into my place. On turning to take the seat behind my

seat by the window, and which they had left, and before
either the man's wife or myself had time to prevent her,
a resolute, hard-browed, middle-aged, bony female,
neither fat nor fair, but perhaps forty, plumped her
"round bones" with emphasis into it, stuck her sharp
features close to the window, and scowled on us with a
triumphant defiance. The lame man's wife, considerably
"riled" at this, began an altercation with this defiant
female, and insisted that I should have the seat; but I
bade the man's wife to leave her alone, expressing a mild
idea in the ear of this mischief-making Eve that if she
had been in truth as much of a man as she looked to be,
in that case I would have pulled her out of the seat to
which she was really not entitled. She only scowled defiance at us both, when, at the wish of the lame man's
wife, I sat by the side of her husband at my original
window, while she took up her position behind us in the
pocket of her foe, and made the journey to the bony
woman, no doubt, a very pleasant one, for I could hear
the broadsides she fired into her until I fell asleep.

When daylight broke through that most dreary night,
to my regret (as we were ascending the wooded hills to
Altoona) I found that a thick fog impeded the view, and
absolutely confined it to the mere width of the rail.
Terribly behind time, and terribly uncomfortable, stopping at every single hut we passed, and putting down
the lame man at one of them (I doubt if this could have
been safely done without the support I gave him), and
passing over two or three good-looking trout streams—or
it might have been the same stream over again—at eight
in the morning instead of six, we arrived at Altoona, when
on our appearance a man rushed on the platform sounding a gong, which I knew to be a signal for breakfast.

Having descended and made some inquiries respecting my baggage, the baggage-master coolly informed me " I had got none," when, on my producing the baggage vouchers that had been given me on entering the boat to cross over to Philadelphia, he explained that " I ought to have given them up to the omnibus man, who took me from the ferry to the hotel to supper." To this I replied, " I saw no man and no omnibus, and that I had not been asked for my tickets, as I had chosen to adopt the freedom of his country, and to select a pedestrian conveyance to where wanted to go." " I reckon you should have ridden, then, and then your luggage would not have been left behind."

Having paid this fellow eight dollars for my dogs (my dogs, in going to St. Louis, cost me more than my own individual travel), I found there was nothing left for it but to quit the train, or break the journey, and stay at the hotel at Altoona and telegraph for my baggage to be sent to me there. Here at last I found a civil gentleman, presiding over the telegraphic department of Adams' Express, who, on seeing my friend's (Mr Hoey) letter, and hearing my story and the extortionate way I had been made to pay for my dogs, at once put the telegraph in motion, recovered my baggage, and arranged that my dogs should be taken with me when I proceeded on my journey as far as Pittsburgh for nothing. The hotel at which I found myself was a very large one, and managed by an excellent and civil gentleman, who studied my comforts, arranged a place for my dogs, and made me comparatively happy. He was himself a sportsman, and I agreed to stay a couple of days and accompany him in search of any game that might be in or about the woods or fields of the Alleghany forests ; and in this resolution to sport we were joined by Barnett, a very excellent sportsman and naturalist, up to every species of woodcraft, a collector of birds'-skins, and a good shot. In his calling he was attached to the railway traffic department at Pittsburgh, and from him I also received the utmost attention and civility.

Here, at Altoona, I had time to consider the boasted comforts of the American railway. Some of my friends on board the " Africa," in the voyage across the Atlantic, had assured me of its infinite superiority to the English lines, thus :

" Yes, sir; you should see our railway carriages. Guess you'll be surprised. None of your little carriages, shut up, but each car like a great long passage; a path up the middle ; yes, sir ; and seats each side, with a door at each end. Yes, sir ; at night—yes, sir—fine beds, so comfortable—sleep all the way; and if there are three cars, you can walk out of one into the other, and do as you like ; yes, sir."

Oh ! after I had experienced the travel of these boasted trains, how I longed for the cleanliness and privacy, and civility and choice of society, on the railways of Old England. The American trains are filthy, their floors not only always in a most disgusting condition, but the door at either end permits such a thorough draught right through, and the citizens of the United States have such a perpetual desire to open and shut them, that any man used to comfort is sure to catch the ear-ache. And, oh ! as to the state of the ladies' dresses ! The hems of *those* beautiful white garments, that ought to be so snowy as to invite the lips of man, are stained three inches high with the filthy tobacco-juice, which it is impossible for them to escape.

At Altoona, then, I took up my quarters for a couple of days, and, for the first time during my visit to the United States of America, my baggage at that time not having overtaken me, I borrowed a serviceable-looking double shot-gun of mine host, ordered my setter and retriever to be got ready, and prepared for sport.

On the morning after my arrival at Altoona, and after a very good bed, splendid cold bath, and a breakfast very well waited on, accompanied by a lad from the hotel, and my two favourite dogs, Brutus and Chance, I set forth to wander in the Land of Liberty wheresoever the leg listeth. In the cloudless sky the sun shone with intensity, and to escape the haunts of men I had to walk up the railway, through the cottages of the town—children, pigs, and dogs playing on the line between the rails, and not a fence on either side to keep off the cattle. Some of the cornfields abutting on the line were protected with the usual zigzag pile of wooden rails, so put up for the want of nails; but these fields were not so much fenced with an idea to the retention of their own cattle as to keep out the cattle of others straying upon the line—a waif-and-stray custom which suggested to the go-a-head mind of the locomotive maker or engineer that admirable appendage to the front of the engine in America, called the "cow-catcher," which so far surpasses the two irons that are supposed to clear the rail for the wheels in England. The "cow-catcher" is a strong iron fence, or set of bars, springing out from the engine in front of both fore wheels, and projecting in a sort of continuous half circle in front of the carriage; and the advantage that it has over the English projection is, that when it catches any of the larger animals, or horse or cow, for whom it is specially designed, it

does not cast them, like the simple English bars, in a
sort of game at shuttlecock from one to the other, but
it "fends" them at once clean away, and hence arises
an annual safety of thousands of lives, for without "the
cow-catcher" the destruction of life and property would
be enormous. Pigs and sheep glance from this iron projection as mere trifles, and on one occasion, when, in
pointing to a number of swine right under us, I asked
the conductor what happened if those animals got in the
way, he carelessly replied, "Guess they get cut in two."
"Who suffers," I rejoined, "when you thus destroy property?" "Guess the company smart for it; and it does
good to the settlers, for they get the fair price for their cattle one way, and, as they fix it in another, a precious sight
more than they are worth." "How do they contrive
that?" I asked. "Why, just as this," he replied.
"If these fellows (we're all pretty 'cute in this country)
have a worn-out cow or horse, or one that is 'down' and
can't git up no how, they just puts them on the line to
get 'em killed, and then they swears they were valuable
animals. Giddy sheep always gits afore us a-purpose;
them we don't care for; but now and then, at nights
(when we can't see 'em, and haven't the hint to increase
our speed), them cows and hosses, and it may be an old
bull, do sometimes make a considerable difficulty. Yes,
sir; that's a fact!"

In my way to the fields and woods, and in passing the
thistles which grow on the sides of the rails, I saw the
goldfinches of America in their natural pursuits, with
much the same "twitter," manner of clinging to the
thistles, and jerking flight, but with brighter and with
more diffused and golden plumage than ours; and I
have brought home a few of their skins for the decora-

tion of ladies' hats. The introduction of these beautiful little birds may well come under the notice of the Acclimatisaton Society. It was not in my power to preserve many of the skins of birds other than the blue robin, the woodpecker, and some others, inasmuch as at that period of the year the moult was not sufficiently set, and I had to throw some of my skins away and replace them with some of the same sort, in better preservation, that I procured on my way home from those excellent naturalists, Mr James Booth, of Niagara Falls, Canada West, and Mr William Galbraith, of the Broadway, New York.

Having left the town of Altoona some way behind me, I gladly climbed over some rails into a stubble-field, that had been Indian corn, shook the dust from my heels, rustled among the beautiful lying for game, or thick wild growth of weeds, which the American farmers seem to take very little pains to be rid of, gave the office to Chance, and prepared myself for a shot. Field after field, stubble, wild, weedy, grassy ground, and good crops of standing clover, were all ranged by Chance in vain, when, seeing that the expectation of any game was a farce, I amused myself by watching the dilemma of the cautious old dog when he came on the golden-breasted meadow lark of America, the size of one of our thrushes, and could not determine by his nose whether he ought to point it or not. He evidently thought he was in a strange land of strange game, for he made doubtful pauses on all sorts of birds which were strange to him, and then looked back at me to ask my opinion; but he very soon found out that they were not worth his notice, and he ranged among them with his accustomed freedom. In the clover-fields I was much struck with the large size and beauty of some of the butterflies,

in appearance and method of flight so very different
from ours; instead of that sort of jerking or insane
method of progression which our ephemeral insects have,
the larger insects in America glided over the sweet blos-
soming clover, or settled as if they were birds. As there
was no sign of game of any sort whatever, I sat down
under the shade of an overhanging wood, fenced from
cultivation by a high zigzag rail, and called on the industrious and persevering Chance to cease his abortive
labour. This he instantly did, as well as reluctantly,
though the thermometer at noon in the shade stood at
75, and in the sun at 110 degrees. On sitting down, I
asked the lad who was with me if the spot was safe on
account of venomous snakes, when, on looking at the
ground, he replied, " Oh, quite ; you never find a snake
where the hogs have lately been, and they have been
basking here." I found, on subsequent inquiry, that pigs
and deer are reputed to destroy snakes with their fore feet,
and that the poison takes no effect on swine, neither in
a bite nor when swallowed, for the hogs are said to eat
snakes. During my travels in America I particularly
turned my attention to the consequences of poison from
" rattlesnake " or " copperhead," and found that the Indians applied the root of a wild blue flower termed " the
snake-root," growing all over the plains, in the way of
antidote. I believe there are two or three plants, an application of the berry or root from which is deemed beneficial in the case of snake bites ; but the remedy which is
said never to have been known to fail in regard to rattlesnakes of from three to four feet long is what is termed a
skinful of neat whiskey; and when the poison of the
snake and the spirit are in antagonistic action in the human system, it is asserted that a man may drink at a

draught a pint or more of whiskey without being drunk.
If the rattlesnake is old, and has attained to the length of
five feet, and arrived at the full force of poisonous secre-
tion, the natives say that not even whiskey will save the
life of the patient, but that, if fairly bitten, he must die.

My meditations on butterflies and snakes were for a
moment interrupted by the swift flight of a passing bird,
and, catching up my gun, I killed that beautiful little
bird, the sparrow-hawk of America; in plumage something like our female kestrel-hawk, but more pretty and
much less, being about the size of the English hobby.
Brutus brought the bird to me, and my attendant seemed
deeply impressed with the goodness of the shot from a
gun that had not been previously to my shoulder.

Having reloaded my gun, I again commenced conversation with my attendant as to all he ever knew or had
heard of in regard to the bites of snakes, and he was the
first of many men I questioned who I found had never
heard of the death of a cow, calf, ox, or bull from the
bite of a venomous reptile; nor could I afterwards discover that any native, white or red, ever suspected while
out on the plains that he met with the carcass of a bison
or buffalo that had died from such effects. Men there
were that assured me that horses and mules had died
from the bites of rattlesnakes; but as to kine, though
they had known swelled faces and swelled legs from what
they suspected was the bite of a snake, they had never
known one to die. The question then arises—and to it
I would invite the notice of my friend Mr Buckland, of
the Second Life Guards—whether or not the vaccine
matter from the cow may not contain a specific against
this poison of the snake to the extent to which it certainly
annuls the virulence of the small-pox, and probably

thwarts the fatal nature in the distemper of the dog? Many curious stories were told me of the bites of snakes in quarters where I have no reason to suspect untruth, and I will here relate two of them.

A young healthy American was out one sunny day, and saw basking in the road before him a rattlesnake. Having "snake boots" on, of very thick material, reaching to the knee (in America they make these boots very well), he, without hesitation, jumped with his heels upon the reptile, but in crushing him he missed his first attempt at the head, and was aware that the snake struck out, and with his mouth hit the strong upper leather on the great toe of his foot; but the man deemed the attempt to wound him abortive, as he felt no pain, and in consequence he took no more heed of the matter. Days and weeks progressed, when, at the end of a considerable time, the man was aware of irritation on the ball of the foot, but the time that had intervened between his combat with the snake and the appearance of this irritation put all recurrence to the probable cause of the ailment out of the question, and he applied such remedies as he deemed the case required. He sickened, however, and eventually died, when his brother, being his heir, not only "jumped into his shoes," but into the identical boots in which the deceased had killed the rattlesnake, and wore them in supposed safety for a considerable time. At the end of some weeks, however, the heir to the boots found that he, too, was, if in a legal, still in an uncomfortable position, and that he, too, had an inflammatory action in the ball of the foot, but again no recurrence was made to the snake. Probably the brother had never heard of that transaction, and the remedies he applied were not those likely to guard against the poison of a deadly reptile. This man also sickened and died, when, as his ailment had been more closely watched, suspicions were entertained of snake poison, the manner of the death of his predecessor also referred to, and in course of public conversation the fact of a combat with a rattlesnake was discovered and the boot minutely examined.

On the most close inspection it was then ascertained that the snake had not only bitten, but had absolutely broken off a fang or holder in the leather, and that the very extreme point of the tooth had attained to the inside and remained there, not protruding sufficiently to give immediate pain, but by friction to cause a slight abrasion of the skin, and thus the poison was infused. This story, if quite correct in all its points, rather shakes the supposition that the tooth itself without its bag of venom is innocuous; but then there might have been sufficient poison inserted into the leather with the tooth by constant friction to have affected the system.

The second case of the bite of a rattlesnake which was brought within my notice was that of an old woman who kept fowls, one of whom had been in the habit of nesting in an old stump of a tree, into which the old woman had to insert her arm whenever she went to take the eggs. On paying one of her daily visits to the nest, in withdrawing her arm she not only felt but saw that, as she supposed, she had very slightly scratched her arm against some point of the jagged wood. The wound was scarcely perceptible, and she took no further notice of it till she found herself in pain, and that the skin was inflamed; and then she applied some common remedy, but not the one that snake poison would have required, and the old woman died.

At her death, the symptoms attending her illness having raised some suspicions, her friends repaired to the old stump of the tree, and on breaking it open for the purpose of examination, curled up in the hen's nest they found a rattlesnake, and killed it. This also is curious in regard to the hen and the snake, and would rather tend to raise the belief that either the snake will not bite a fowl, or that the venom to a bird of that class is not fatal.

Having learned from my attendant all he had ever heard or known of rattlesnakes and copperheads, I cast Chance off again in search of game, when, in crossing a small rill of water, Brutus, on going to drink, flushed the little water-rail, so common all over America. The bird is about the same size as that which we have in England, but rather shorter in its make ; and all the specimens that I killed were slightly speckled with white. In this instance the rail only flew a couple of yards, and dropped before I could shoot, and nothing that myself or my dogs could do would put the bird again to the wing ; so, giving up all idea of game in the fields, I betook myself to a rather large wood, fenced with the usual zigzag rail, to beat for a woodcock or a rabbit, the one being as scarce as the other. I had not been long thus occupied when I saw something running in the midst of the moss-grown limbs of a huge prostrated tree, and I lost sight of it behind the butt of a tree which was standing. I felt sure it was not a young rabbit, and had suspicions of a squirrel, when, on reaching the spot where the animal disappeared, I found a hole in the decayed tree, into which the creature, whatever it was, had made its escape. On my attendant coming up, he pronounced the animal of chase to be a ground squirrel ; so, seeing that the trees all round me were in a fallen or falling and decaying and

neglected state, all my English notions of private property
and the value of timber became more acclimatised to the
realms in which I then found myself, and to my attendant's delight and increased notions of my proficiency as
a "hunter" (in America the term for all sport with a
gun, whether after animal or bird, is "hunting"), I bade
him collect moss, dry sticks, and leaves, and set fire to
the tree. I soon found that he was an adept at this sort
of sylvan proceeding, and he speedily, with his pocket
implements for lighting his pipe, got up a flame, directing it as much as lay in his power into the small hollow
of the tree. During this proceeding Brutus and Chance
sat by in considerable curiosity, Brutus having made up
his mind that, if not a rabbit, there must be something
in the hole that he had better be ready to catch; so, with
my gun resting against a tree behind me, and an eye
occasionally to the boughs above of the tree on fire, in
case there should be a "bolt-hole," I tended the interests
of the incendiary, and waited the result. The fire having
continued some time, and no smoke making an exit from
any other portion of the tree, I informed my man that
the squirrel must be dead, and bade him clear away the
fire, and remove some of the charred bark. He had not
obeyed this order a second before he exclaimed, " Here
he is!" when, on taking his place, I perceived a little
tail, which, on laying hold of, I found to have been
burned, and the skin came off in my hand; removing a
little more of the bark I then possessed myself of the
body of the beautiful little animal, which had been suffo-
cated, and admired the striped and brilliant hues of its
glossy skin. From the loss of the tail it was not a perfect
specimen, but that and the sparrow-hawk being the two
first things I had killed, on my return to the hotel I pre-

served their skins. After this adventure I saw another of these pretty little animals, but contented myself with observing as much of his actions to escape me as I could, and left him to enjoy his life till he met with some more bloodthirsty " hunter."

There being really no game in this portion of the country —there never is in America in the immediate vicinity of a populous place, unless a large river, or some other obstacle, cuts across the path of the Sabbath-breaker and thwarts his shooting propensities, Sunday being the day when all are idle and hardly any at church—I gave myself up to the contemplation of the quiet scenery during the splendid and airless afternoon which followed, and amused myself with thinking that perhaps the old bear and her two cubs, which had been killed in those forests but a few days prior to my arrival, might perhaps have been on the spot where I was. How wild, hushed, and strangely beautiful those primeval forests were, and what an air of grandeur, as well as desolation, that sylvan scene afforded! Strewn upon the ground lay those large trees that had once been mighty monarchs of the forest, not even honoured in their decay by the passing notice of the poorest of the adjacent settlers seeking fuel. At intervals, the same class of dead kings of former verdure, victims of some Indian fire, who had not yet fallen from decay, or been torn up or stricken down by tempests of wind or lightning, so prevalent in America, stretched their bleached arms above the surrounding verdure, as if seeking annihilation by the same element from heaven which man in design or wantonness had blasted them with on earth; pale, bare, and glittering in their ghastly hue, they offered a strong and startling contrast to the invisible green beneath them, deepening into shade by the decline of the setting sun, which now began to throw the long last eastward shadows, those ever-attendant satellites of an unclouded and a daily reign. In the silence of these grand solitudes how interesting was it to me to hear their stillness occasionally broken by the cries of birds speaking to me, not as in England, but in voices that gave me not their name! Among them, though, I recognized the shrill and occasionally hoarse cries of the American crow, sometimes reminding me of the English jackdaw, and at others of the carrion crow. The woodpeckers I traced by their stationary cry from the stems of trees; there were no birds then in song.

As I sat on one of these fallen trees, the hum of what seemed to me a very large gnat reminded me that I was in America, and, according to the intelligence afforded by my English friends, about to be subjected to insect appetite and persecution. I sat still to await the approach of this novel foe, who was so kind as to herald his advance by a very distinct trumpet sound, and on he came, a giant gnat, about twice the size of an English one (the mosquito is nothing more), when, on attempting to settle on my cheek, he met with the palm of my hand from his rear, and his body was soon given up for inspection. I sat some little time after this, when a slight whistling of wings, coming from behind me and over the trees, made me snatch up my gun from its recumbent position across my knees, and as it came to the level and discharge at the same moment, I knew what I shot at, and heard my man exclaim, "Pigeons!" It was a very difficult snap-shot, and though nothing fell, something told me that the aim was correct, so having loaded I followed the line of the seven wild pigeons. In a very short distance the sudden flap of a wing among boughs told me

that a wounded pigeon was darting away, whose further
flight I at once arrested, and in a moment after, and
while Chance sat down marking the place where the bird
fell, Brutus came rejoicingly back with the pigeon in
his mouth—a fine plump bird, not so large as our quiet,
and more in shape and feather like our turtle dove, though
of greater size. The tail much longer than the tail of
English pigeons, and the outside feathers being the
longest, assimilating to the forked tail. Having thus
thoroughly impressed my companion with the shooting
of an Englishman, as well as with the docility of his dogs, I
returned with an appetite to mine inn, and found that
the host had attended to my suggestions, and provided
something a little more palatable, but I fear not anything
like so wholesome as tea, for me to drink at dinner.
Hear it, ye Tam O'Shanters of the United Kingdom
of Great Britain and Ireland, in all cases the rail-
way-side inns are forbidden to sell spirituous liquors,
I suppose because, a free people having no efficient
means of restraint either in themselves or through police
constables, and there being no second-class carriages for
Boh-hoys, whose immorality might come out strong through
the genial spirit of whiskey, it would not be safe to let
conductors, or some of the conducted, have free access
to an enemy that steals away the brains.

When I went forth in the morning mine host assured
me that he would get me some trout for dinner, which,
according to his account, were plentiful in the adjacent
streams that descended from the Alleghany mountains;
but, like many another pledge on sporting matters in
the United States, I found perhaps the wish of my informant to be the father of his promise, and I had no
fish for dinner.

CHAPTER V.

ON the following day, Saturday, it was arranged that
mine host, myself, and Mr Burnet, should start by train,
and, in easy distance, reach a portion of the forest where
game was said to abound. Brutus and Chance were to
accompany me, and mine host was to take with him his
very good-looking setter, while George Bromfield, in
case of a wounded deer, was to have my lurcher Bar in a
leash. We started about ten in the morning of a splendid
but intensely hot day, and on reaching a few huts near
the residence of that most hospitable and intelligent
gentleman, Major B. F. Bell, of Bell's Mill, we deposited our basket of luncheon at the little station, and
proceeded on foot up the rails for some distance, and
then broke away into the cornfields and forest. Having

traversed a considerable quantity of cultivated land without seeing anything more in the shape of game than one water-rail, and not relishing the amusement of mine host, who delightedly shot any meadow lark that rose, and shot them very well, I went up to the first farmhouse I saw, and, on finding a very civil and communicative agriculturist there, asked him if he could direct me to any game, or if there was such a thing in his vicinity, or if he had seen any since the last spring. He laughed outright at the bare idea of game, and assured me that, as far as his settlement went (and it was rather considerable), his lands were quite as bare of game as the idea itself that I had so recently broached, and that though he would not undertake to say that if I beat all the woods I might not see such a thing as the bird they call the pheasant, still he " considerably guessed that I should have some difficulty in finding anything like one." On receiving this assurance, and being in no haste to weary my willing dogs under such a sun, and in severe ground, on a wildgoose chase, I entered with my friend on general conversation, and, among other things, I asked him as to his ever having crossed his stock with the buffalo or bison of the plains, or known it experimented on by others. He replied that he had had one cross himself, and that she was a very good milch cow, and very hardy; that he believed that the cross was not a mule, but would breed again; the great objection to it, however, was, that from the large size of the shoulders or hump of the offspring from the buffalo or bison bull, the domestic mother was always in danger of a very hard " time," and frequently died in giving birth to the calf. At the farm to which I now immediately refer there was a barn, but usually in the cabins of farmers throughout the scene of my travels the very last thing they think of is a barn, or a farm-yard or garden. They leave their corn out all the winter in shocks,—I speak of the corn we should call Indian corn, but which they call corn; there is a less description of similar grain which they call Indian corn,—and only bring it in as they want it. The obvious consequence of this slovenly state of things is, that they lose half the crop through the depredations of wild animals—rabbits, rats, mice, birds, their own infracting pigs, in some places deer, and from the weather.

This, however, seems to occasion them no uneasiness. They have squatted in this coveted but overrated independence; they have bacon to exist on and corn, and corn enough besides to procure them the deities of their lives (many of them have no other deity)—tobacco and whiskey; and so long as they can smoke, chew, and splice their shattered constitutions with frequent drams, assailed within and without by intemperance, and ague, and fever, as the consequences of climate and decayed vegetation, they care not for the unhoused corn, the safety of which would occasion them too much trouble, even if they turned it into dollars.

I know not anything that is more strange to an eye used to the cornfields of England than the sight of lands in corn, as I saw them in those distant parts of the United States where I made these observations. Let my agricultural friends and countrymen fancy a large field, fenced with a zigzag flight of rails, placed to rest on each other at sharp angles, without a nail or a post to hold them together, and high enough to keep out deer; and then, in an inclosure for corn thus made, let them imagine at intervals huge tall dead forest-trees, standing up like the ghosts of vegetation in that field, with the stumps of others from four to five feet high, also erect, and all of them of a most intense whiteness, the plough having done its office between these monuments of idleness, and scratched up an intervening crop of corn. The way that this is brought about is from the farmer or settler (who in better-regulated countries would have been the labourer of other men possessed of capital, instead of an abortive agricultural squatter on his own hook) having no hands but his own to aid him in his toil, the task of a thorough clearance from wood of his lands would have been too much for him. Hence he grubs up the superficial shrubs and bushes, and makes them into a fire at the root of the larger forest trees, and he sets the whole thing in a blaze at a dry time, and trusts thus to kill the trees. Those that escape death from the effects of the conflagration he " girdles," that is, he barks a foot or more of their entire circumference, and thus stops the circulation of the sap by which they live. By fire and " girdling " he thus kills all trees within his inclosure, and then leaves the storms to prostrate them as they gradually decay, when of course he removes their bulk for the purposes of fuel. I have heard of a machine which the better class of these strange farmers use for the riddance of their land from old stumps of trees, which they call " the tooth-drawer," I believe propelled by oxen or horses ; there is a lever to it, which extracts the bole of the tree, roots and all ; but I never saw the machine in operation. Pumpkins and melons, all excessively fine, and sweet potatoes, which I think detestable, seem to me to be the only garden produce they care about. They all keep fowls of the common sort, very largely mingled with the Cochin China breed, and therefore the eggs which you procure at all the cabins or agricultural huts are infinitely better than their chickens.

My agricultural discourse with the farmer being come to an end, we again proceeded in the search of the phantom game; but this time we ascended into the woods, and I had a shot at what might have been a woodcock, at least I thought so. I did not kill it, however, and could not flush it again. By the side of a little stream in the forest, Burnet, who I saw was well up to any sort of woodcraft, padded a "skunk" and a racoon, and we found one of those little land turtles, which I afterwards met with in great quantities on some parts of the plains; and we heard the large woodpecker with the crimson head, but could not get a shot at him. Tired, as well as disappointed by our want of sport, we then proceeded to regain the railway station, and, on arriving there, ascertained that there would not be another train for some time. On hearing this the host of my hotel conducted me to the residence of Major Bell, who, with the greatest kindness and hospitality, set before us viands that added zest to already existing appetite, and I saw that his house and mill were cosily situated by a pretty little trout stream.

While thus having luncheon with him, Major Bell showed me several very curious specimens of geology from the limestone rock and other strata of the Juniata valley of Pennsylvania, and presented me with some arrow-heads in stone. Among his collection was a flat stone, containing the most perfect impression of a shoal of tadpoles, all, at the time of their destruction, swimming together in one direction, and so perfect was this impression that even the turn or wriggle of every tail was as minutely depicted as if the owners of the tails had that moment been before you. On my taking leave of my hospitable entertainer, he presented me with " The History of the Early Settlement of the Juniata Valley," by U. T. Jones, and dedicated to him by the author. The work is very nicely got up; but I regret to have met with in its pages a bitterness of expression towards " the old country " which ought never to appear, and which, I trust, by this time has been entirely obliterated in " the new " by the consideration of the very obvious interests which should bind the two nations together.

On leaving the hospitable residence of Major Bell we repaired to the railway station in the woods, there to await our train, and in the red glare of the evening sky of that most charming sunset I became intensely amused by the sudden appearance of a hundred or more of the common night-hawk or goat-sucker (such as we have in England), which, rising high in air, occupied themselves with a manner of flight very like that of swallows in catching insects; and their graceful evolutions were confined to the region immediately around and above us. Burnet and myself having differed or doubted as to the height of their flight, we waited for those that came immediately over our heads, and fired several shots at them. My luggage having arrived the night before, I had then in my hand one of my old favourite John Manton guns of the eleven gauge, and I still think that if I had had a cartridge of No. 3 shot I could have killed some of these birds, even at the height at which they were; but with loose shot, and even with Burnet's enormous wild-fowl gun, we failed to bring one down. The train then arrived and we returned to Altoona, and having ordered George to clean my gun, for the first time I became aware that a traveller in America bent on sport ought always to carry tow as well as sweet oil with him. At Altoona there was no tow to be procured, and

at the smaller places further away from civilization there
never was any sweet oil.

Having remunerated the black servant (being a slave, of
course a good one as compared to any American) who had
assisted George Bromfield in the care of my dogs (George
had slept in a loft with them), I packed up my things, and
prepared to continue my journey to St Louis. I must not
forget that during my short stay at Altoona I met several
American gentlemen who evinced towards me the greatest
kindness and courtesy, and who were loud in their condemnation of the extortion practised on me by the Boh-
hoys of the baggage-cars on the trains. In walking down
the street while I was there I also met an English face
that had considerably brightened up as I approached, and
I was civilly and rejoicingly accosted by a countryman
of mine, offering me all congratulations on my visit to
America. On my expressing my want of recollection as
to *who* he was, he replied, " The last time I saw you, sir,
you gave me a dinner and a bed at Beacon Lodge, when
I came over but just in time to attend the funeral of poor
Mary, your old servant and housemaid, who died at High-
cliff, in the then service of Lord Stuart de Rothesay. If
my services can be available to you I shall thankfully
render them, if but in the shape of acknowledgment for
your kindness to me and to mine. I was the nearest relation that poor Mary had." Thanking him for all his
kind expressions, I declined his proffered services, and
I must now carry the reader over miles of rail in the
direction of St Louis.

The train, as usual, was late in its arrival and departure from Altoona; but when it came I was charmed to
find that it was not only kept in a greater state of clean-

liness than any I had been in before, but that in charge
of the van there was (by the greatest good luck) a really
civil and respectable baggage-master, who, at the request
of Adams's express telegraphic agent, agreed to take my
dogs as far as Pittsburgh for nothing. In this train, and
further on, I met with several gentlemen and ladies of
the upper classes, well informed in all matters, most
agreeable and delightfully good-natured, and, if I am not
incorrect in my recollection, I cannot help thinking that
among them (betwixt Pittsburgh and Cincinnati) I met
with Mr and Mrs Sullivant, going to their residence at
Homer, Champlain County, Illinois, who most kindly invited me to pay them a visit. The pencil note of the address, made at the moment, has become so nearly effaced
that it is with much difficulty that I can read it; but I
am fain not to pass over, from apparent neglect, some of
the most agreeable and kind acquaintances made by me
during my visit to the United States. We began, then,
the gradual ascent of the Alleghany Mountains on as fine
a day as any traveller could desire to be out in, and the
scenery was perfection! To have a better view, I left the
inside of the train and stood out upon the little landing-
place behind, where the breaksman stands to regulate the
speed, and thence holding on to the hand-rail, I could
see all around me, and make a note of all I desired to remember. On either side, the high hills arose, clothed in
timber to their summits, the foliage occasionally varied
by the jutting into sight of rocks and crags. Beneath
these crags might have been caverns for the dens of bears
who still haunt these forests, though their race is nearly
run, on account of the gradual approach of man. At intervals the bleached ghosts of mighty trees, similar to

those that I had before noticed in the vicinity of Altoona,
stood forth above all others, and stretched their pallid
arms to the sky. The autumnal tints had begun to break
or flash from the midst of the varied hues of the oak—
the oak called the "black Jack"—the chestnuts, the hic-
cory, the beech, the ash, the sycamore, the hazel, and
from other trees as well as from the many growths of
shrubs and bushes, the wild vine and weeds, which make
the underwood of the American forests so varied in hue
and so very difficult to force through.   I noticed all
these trees that I have named; among the foliage there
were large and small patches of the most vivid scarlet,
and scarlets, and vermilions, and yellows of varied grades,
that were intensely charming and picturesque, while, as
we proceeded through cuttings in the sides of the hills,
the iron ore "cropped out" occasionally in the greatest
profusion, showing me the mineral richness of those
primeval mountains and woods, and of their as yet un-
explored treasures, prone to the hand of future generations.   In the air, that scavenger whose devouring inclination towards carrion is so well recognized in the United
States, whose beak I fear is often the veil to violent
death, and for the protection of whom Congress has made
a law—the turkey buzzard, soared in a graceful flight,
similar to that of hawks or kites, and at first I took them
for birds of that species.   Among the Americans who on
that day were with me there seemed to be a total ignorance of ornithology, and want of interest in that as well
as in the names of forest trees or shrubs, and not one
soul of them could tell me on these heads more than I
knew or guessed myself.   I have seen this remarkable
want of interest in things pertaining not to dollars,

widely spread through certain classes in the United States, and I attribute it to the simply mercantile spirit that completely engrosses the mind of many a good man.

We passed Cresson House, an hotel open only in summer, fifteen miles above Altoona, and from what I saw, and also from what was told me (though I am far from putting faith in hear-say), I can conceive that hotel being a very healthy and a rational and an amusing place in the heat of summer, both for the gun of the ornithologist or the hand-net of the entomologist, as well as for the rod and line of the fisherman, as there must be trout in the mountain streams. We now approached that curve in the line I was so anxious to see, and I do not know that I can describe it better than in likening its shape to that of the fore-foot of a donkey. A little narrow valley bites into the mountains, and when you are railing along one side of that valley, you see your road within rifle-shot on the contrary side, and the train, to keep on its given way, absolutely has to describe a very limited half circle at the end of the little valley in the side of the hill, with a great depth beneath it, and an inaccessible height above.

I stood outside the carriage and thought of what would happen if some reckless go-ahead engine-driver, supposed by my countrymen to be so prevalent in America, should get one drop too much excited by whiskey from a secret flask—a fault to which a man even in the best-regulated family may occasionally be prone! A very decreased rate of speed was all that kept us from destruction. For a length of way we wound up and then down through this magnificent scenery, stopping as

usual at every little hut of the free citizens, to show, I
suppose, that the city and the cabin were alike considered; but to the very great loss of time and, I should
deem, to the waste of steam and fuel, for the merchandise delivered or the passengers put down were often
none at all.   During my progress from Altoona and
Pittsburgh to Cincinnati and Staubenville the train con-
tinually and for miles ran through the primeval forests
as well as cultivated lands, and, when not ascending or
descending the Alleghany Mountains, over a very flat
line of wooded country, and there being no fence on
either side the rail, the danger of collision with cattle,
which were straying in all directions through the forest,
was imminent. We all had, however, or seemed to
have, a lively faith in the "cow-catcher," and the greatest contempt for the lives of pigs, sheep, men, women,
and children, or the obstruction and distress that their
dead bodies might occasion; and though some of the
passengers had distorted limbs, or limbs weakened by
fractures from accidents on that very line of rail, nothing
seemed to keep us from being very jolly, nor to stint the
good humour of the company.

" *Does* the cow-catcher," I asked, " *always* ' cant' the
beef on one side?" "Yes," replied the conductor,
" guess it does, but I *have* known a cow canted upwards
and carried on the top of the catcher for some distance;
however, not long ago on our line a bull gin us a share
of his difficulty, for instead of getting out of the way on
the sound of the whistle, on he came at the top of his
speed, full butt, tail on end, and guess there *was* a con-si-
der-*a*-ble smash, for he was split, the engine knocked off
the rails, and the driver killed. Cows take it easier;

yes, sir, and pork, sir, goes for nothing; yes, sir, that's a fact."

Rattling on in this way, in conversation with my amusing friends, we came to Pittsburgh, and when my dogs were removed from the luggage-van, I observed they were suffering terribly from heat, and begged of all I saw to tell me where I could get them some water. Burnet, who had been out shooting, or in an attempt to shoot, with me while at Altoona, was there rendering me all the attention he could, but he too failed in procuring water. At last, by the offer of the silver coin that does duty for an English shilling, a spawn of a Boh-hoy, in the shape of a dirty little child, was roused from his listless idleness; he had scorned all allusion to a "dime," the amount of which would have moved bigger boys in England to walk a few yards, and at last he brought my dogs some water. This, the obtaining water for my dogs, left me but little time to swallow (eating was out of the question) some of the coarse food that railway house of entertainment afforded, when, having struck a bargain with the baggage-master from Pittsburgh as far as Staubenville, for the conveyance of my dogs—the amount charged, I think through the intervention of Burnet, being but two dollars and a half—I was putting them into the van when the following rudeness was offered me. A great dirty, bearded, crumple-brimmed, high-crowned hatted fellow jostled against me with his trunk, and very nearly threw it on Druid's back. So rude and offensive was his manner, and so needlessly rough his method, that I lost all patience, and turned sharply round on him—he was rather behind me—and said, "Gently, sir, I'm not at all disposed to put up with insult, nor to have the limbs of my dogs thus wantonly endangered, so no more of *that !*" The fellow scowled at me from head to foot, I suppose to see what sort of a customer I might be in a " rough and tumble difficulty," said nothing, and, I take it, possessing no more property than his one trunk contained, did not desire to place himself again in an offensive attitude.

CHAPTER VI.

WHEN we left Pittsburgh, that coal-blackened proto-
type in hue of the English Newcastle, I had time to
reflect on the vale of Juniata, and the very remarkable
spring said to exist in that beautiful locality on the right
bank of the river, about seven miles below Hollidays-
burgh, which I regretted much that I had not time to
see. It is said that the spring is of the purest limestone
water, and that it regularly ebbs and flows by day and
night. At one time the spring is full, at another empty,
so that the basin only, where the water was, remains. A
rumbling noise as of water is then heard higher up the hill,
immediately above, and gradually the spring fills, and thus
it continues to ebb and flow for ever and for ever. Through
inquiries made by me I could not ascertain that its

ebbing and flowing were in any way governed as the
tides of the sea ; but it seemed to me, from what I could
learn, that the water receded or advanced as if from
some volcanic commotion or rocking within the hill ;
but as to whether the water was warm or cold, or
as to the stated time of its flux and reflux, my
informants could not afford me the desired information.

Soon after quitting this black diamond in the star-
spangled hemisphere of the United States, Pittsburgh,
standing as it does on the Alleghany river, we proceeded
through a flat line of country, of settlement and forest,
which, in spite of monotony, charmed me with its wildness, and filled the sportsman's mind with anticipations
of the chase.   The railway then occasionally passed over
or touched on the Ohio river, at least I think it was that
river or its tributaries, and there was nothing particularly
remarkable during the journey, save the constant danger
resulting from there being no fence to the line, and the
presence of cattle, allowed to stray where they pleased.
Having gone ahead the whole day, and swallowed hasty
repasts, doing duty for luncheon or dinner, with nothing
to drink but tea or coffee, we arrived at Staubenville,
where we were delayed two hours.   If my memory serves
me, for my notes as to this portion of my travels were
subsequently somewhat deranged by the rough usages
of the desert, we arrived at Staubenville in the night, and,
having got out with my dogs on the platform, I bought
up all the milk the vendor of small refreshments had, and
the greater part of his rolls, and fed them there and then,
giving them just exercise enough to stretch their legs.
Having replaced them in the van, I then retired to my
seat in the car, and amused myself with watching the

rats on the platform examining luggage to find out any-thing to eat.

The train for which we had to wait arrived at last, and we proceeded on our road, and during the following day journeyed through fine settlements for grazing cattle, of which I saw plenty, also some sheep, as well as very large numbers of pigs; and, for a wonder, the farmers had fenced their fields on either side, and thus afforded better protection to the lives of those who passed their lands by steam. During the middle portion of this journey, and at a time when we were in sight of the Ohio river, on one side of which was a Slave State, while in agreeable conversation with some ladies and gentlemen, I jocularly remarked on the great disappointment I naturally felt, in passing the larger rivers, not to see romantic and heart-broken "blacks" gracefully seated on "snags" or pic-turesque promontories on the banks of the streams, melo-diously and pathetically singing of innumerable "Mary Blanes" and "Lucy Neals," of whose fond attachments they had been reft by their cruel masters. I told my friends that it was the general belief in England that the scenes on the banks of rivers in America were always deeply interesting, from the perpetual dancing and sing-ing of black men, but that up to the time at which I was then speaking I had met with nothing of the sort.

"Guess you have heard fine stories," replied one of the gentlemen; "I'd like to know—hum—um. Tell us some more."

"Well," I continued, "I assure you that in my coun-try it is the general belief that the negroes, though paint-ed by the anti-slavery humbugs (reader, I detest slavery, but I hate humbugs who deal in falsehood under a pre-tence to serve Heaven, just as much), as the most cruelly treated wretches in the world, do little else than move amiably-inclined white men to tears by the artless but beautiful melodies they are for ever chanting. How is it, then, that those negroes that I have seen—save some at Altoona who helped to take care of my dogs, and also acted as porters to the hotel—not only never sang, but without any exception were the idlest, most ill-looking, discontented, hang-dog ruffians that ever made a man uneasy as to his watch, or arrested the vigilance of a constable ?"

" Oh," exclaimed one of the ladies, " you have as yet, or generally, met but the *free blacks*—they are dreadful. If you want to hear singing, music, and dancing, and to *see the blacks really enjoying themselves*, you should contrive to spend some days in the *Slave States*, and then there might be some truth in your anticipations." " What ! " I cried, " how glad I am, then, my dear lady, to receive this fair assurance from you of a truth which I suspected, or really knew very well, but which all the anti-slavery humbugs in my country have been trying for their own democratic purposes to cry down ; *if* I am to see the negro population *really* ' enjoying themselves,' I must seek them, then, *as slaves*. A pretty good blow in the face this, in regard to the assertions of Mrs Trollope, Mr Dickens, Messrs Bright and Cobden, *et hoc genus omne !* " " Hush, my dear," said one of my kind anti-slavery male friends to, I believe, his wife, he at once perceiving that the truth let out of the bag was somewhat startling to anti-slavery men, and thus unguardedly exposed before " a chiel," as the Scots say, who was " takin' notes," and " in faith would prent them "—" Hush, my dear, you are talking of what you don't understand." The lady looked at me, laughed, and said no more, but

I continued to amuse myself with putting questions to gentlemen, the answers to which completely tended to prove the truth of the female lip. Though these anti-slavery men of course raised that very handy shield of "heavenly intentions and the rights of man," there was not one of them who would have admitted a man of colour, when free, to the courtesies of a dog, or who attempted to gainsay *the fact* that those negroes who were emancipated had degenerated in the social scale of existence, and become generally the very refuse of the very lowest grade of the American population.

I forget the hour at which on Saturday, the 11th of September, we arrived at the large and influential town of Cincinnati, to the principal hotel of which I had telegraphed from Altoona to bespeak a bed for myself and a lock-up room or stable in which my servant and dogs could sleep. Arrived at the hotel, which is a very large one, I entered the ample hall or lounging place in front of the bar, considerably stared at by faces catching glimpses of me in the midst of tobacco-smoke from between the thighs, legs, and feet of the bodies to which those faces belonged, as they balanced themselves in upside down fashion on chairs or benches, rather than sat, which had been arranged for the convenience of customers. There also were the spittoons hopelessly placed against pillars or the sides of walls by despairing Irish housemaids, whose laborious duty it was every morning before daylight to scrub the filthy floors.

On entering the hall, then, all the gentlemen that were there stared at me, as I walked up to the bar, behind which stood what I immediately saw was, no doubt, a military officer of high rank, and I asked of that most dignified-looking volunteer if a telegraphic message had been received from me, regarding my bed, my servant, and my dogs, and I announced my name in full.

The gentleman to whom I spoke turned carelessly round to a companion and asked "if a telegraph message had been received." The one applied to, after some dubious hesitation, said "he thought it had." "Think!" I said, somewhat tartly; "have you not condescended to read it?" "Guess it was something about dogs, then," the man replied. "I dare say we can fix them." "What then," I rejoined, "you have not paid any attention to the request expensively conveyed to you!—not much use, then, in *your* telegraphic wires. Look sharp now, at least, and see if you can 'fix' my man and dogs until Monday in a place where he can sleep with them, and give me the key of my room, as I am dusty and tired, and want a bath." My key, labelled with its number, was then handed to me, and I at once bribed the Irish porter who carried up my dressing-case and tin bath, to bring me some pitchers of cold water. I had but just finished my ablution and nearly dressed myself, when I heard a long fast step striding up the many stairs to my apartment, for they had given me a bed-room at the top of the house, and then came a hurried knock at the door. Admission being given, a tall, good-looking young Englishman stood before me, who, announcing his name as "Easton," and referring me for a knowledge of his father to Strathfieldsaye, told me that having just heard of my arrival he had run up to offer me any kind attention in his power; but before I could thank him he cast a scornful glance round my apartment, and said, "But there's some mistake, you're in the garret; this won't do." He was turning to the door, when I said, "Oh yes, it will suffice; the bed is clean and comfortable,

I shall soon have a harder pillow, so it will do very well."
"No, no," he replied, and without another word left
the room. Shortly after his exit I heard two people
coming up-stairs, and Easton threw open my door, followed by the now civil and excellently well-mannered
gentleman to whom I had first addressed myself behind
his own bar, when Easton presented him to me, and introduced us to each other in due form. Mine host at
once advanced, and, shaking me by the hand and offering his congratulations on my safe arrival, concluded his
gentlemanly address by saying, " Have the goodness,
sir, to follow me. There has been a mistake. Apartments much more suitable to your rank are awaiting you
below."

I followed him, and very soon found myself in a bed-
room and sitting-room nearly equal to an hotel in London. Mine host having left the room, I exclaimed to
Easton, " Now, do tell me what on earth is the meaning
of all this ? I telegraphed in my name in full to this
man, and on my arrival here I gave him my name in
full; but he paid not the slightest attention to the telegraphic message, and when I entered he received me
with a brusqueness and rude manner, much as if I had
insulted him by entering his house to pay for the com-
modities he exchanged for money. Now, you have
turned him into gentility of manner and urbanity itself.
How is this ? "

"Why, thus," Mr Easton replied. " No traveller
through the United States is believed to be what he individually represents himself. He must have a known
man to introduce him. The President himself, when
not personally known, is not paid any attention to unless
he is thus verified; and when Mr Cobden was here, he

always had 'the man in black' to promulgate his importance. So many Englishmen and foreigners have previously represented themselves as lords, dukes, and marquisses, who really were the refuse of the countries from which they had probably fled, that if any real nobleman or approved gentleman now comes, the rule is to distrust him till some one vouches for his identity. The title of 'honourable' here is so common, through the appellation attached on election to Congress, that it commands no attention whatever." Having thanked Mr Easton for his great attention, he left me, saying, that as the place to which George and my dogs were consigned in the basement or lower regions of the hotel was neither cool, well ventilated, nor private, he would, with my leave, conduct them all to a room in the house where he held his official apartments as engineer, employed to lay down the street rails for omnibuses in Cincinnati, and where they would be perfectly safe for so long a time as I chose to stay. He left me with a promise to see me again after dinner, and to discuss with me some Badminton mixture which, heated with travel, I resolved to brew.

Having solicited the attention of the chief steward, I then and there, and by way of a happy lesson to him, procured a bottle of claret, two bottles of soda water, a lemon and sugar, a glass of sherry, and sighed because I could not obtain a slice from a fresh cucumber. Then, all having been well iced, and one of the bottles of soda water not mixed with the wine, but kept back for the purpose of being amalgamated with the other ingredients when the whole were put on the table for drinking, I cast myself into a comfortable arm-chair, and prepared myself to enjoy a quiet dinner.

Before I quite rested thus from the fatigue of travelling, I desired the steward to fill himself a glass of Badminton, when, on his expressing satisfaction, I told him
that henceforth and for ever by that mixture he would
be able to distinguish the real Englishman, nobleman,
and gentleman, from the mere English adventurer and
the vulgar ass, who, by aping eccentricity or other
affectation, desired to pass himself off for a great man.
The steward had only to ask the traveller in hot weather if he should make him some "Badminton," when,
if he declined, or expressed ignorance of what the mixture was, then he certainly was no peer. Laughing
at the serious and thankful way in which the steward
seemed resolved to bear himself in regard to these
injunctions, I dismissed him and gave myself up to reflection.

Thus far then on my travels I had learnt a great deal,
and the last thing was that the title of "honourable" being bestowed on men elected by universal suffrage or
by ballot, as practised in America, was not regarded as a
distinction, but passed by as a casualty not deserving of
any peculiar respect whatever.   In this democratic country, by whose political institutions, according to Messrs
Cobden and Bright, Old England is in future to shape her
House of Commons, a title by birth *is* deemed more
worthy of veneration than one temporarily gained by the
universal approval of a giddy multitude! A liberal as I
have ever been, I cannot but congratulate the citizens
of America on this wise opinion, for where intimidation
of the conscientious use of the suffrage is enforced by rifle-
balls and knives, instead of being simply governed by
the desire of a landlord, and where millions in lieu of
dozens dictate at the poll, why, no sensible man of liberal

views can hesitate as to his choice of two evils, and he must prefer the English Parliament, as it was and is, to the Congress of the United States.

The boasted convenience for sleeping through the night in the railway cars was about the greatest mistake that it was possible to imagine. A berth above, about a foot broad, was the upper place ; and two berths, side by side, were the lower ones. I selected the upper one, and my kind fellow-traveller on that occasion, I think Mr Sullivant, secured the two lower ones for himself; so that we avoided the spitting of tobacco, and laughed a good deal at my ejaculations, so hopeless of rest when I climbed into my cage of torture. Attached to all trains there is what is termed a lady's carriage, in which smoking is prohibited ; into this, when it is not full, men of respectable exterior are permitted, and at times men get in there whose looks are decidedly the reverse. Though the railway companies set their faces against two prices, as subversive of democratic principle or no principle, they nevertheless pretend to attach what they call the emigrant car, wherein they will let people travel at second-class price. I believe this carriage to be a vision, or perhaps one for the conveyance of men of colour ; all I know to a certainty is, that men, women, and children can pay an emigrant price, and that though they may be of the lowest of the low, and covered with disease and filth and tatters, at that second-class price they are put into the same cars as those who pay the full price. As utterly a dishonest act as a just man can well imagine.

In England the first-class price keeps a carriage select. But in America there are thus in reality two prices, and the respectable class of passengers are annoyed and disgusted, and their intentions nullified, by having rogues,

thieves, disease, and filth thrust into companionship.   I
believe I may with truth affirm that, at this moment, the
railway system, in its filthy and falsely democratic or
cheap state, does not much remunerate any of the companies ; and in this I have no doubt but that Mr Cobden,
or any other traveller in America, will thoroughly agree
with me.   In short, to invest English money in railways,
or in land, or in any other permanent purchase in the
United States, would not, at this moment, in my opinion be
attended with safety.   I make these suggestions or assertions—let my readers call them what they like—on no vain
imaginations ; and as an instance of the strength of my
position, I need but quote the fact, that many of what may
be called the landed gentry in America, while I was paying a visit to them or otherwise in their society, assured
me that so insecure and prone to aggression from the Boh-
hoys were the rights of property, and so uncertain the
protection of the law, that if they could shift their vested
interests to almost any country under the sun which had
a safer government and greater personal security, they
would prefer living there than in the country where they
were born.   Beautiful as the country of America is, and
in some parts perfect as its climate is, with the richness
of its soil for cultivation, wealth of its mineral productions, and its thousand and one attractions for profit or
sport, terrible it is to see the lower class of inhabitants so
unrestrained and drunk with freedom as to assume that
wealth, birth, education, and talents are no recommendation in the House of Representatives.

I have known landed gentlemen in America to make
attempts to have deer in their parks, after the fashion in
England.   Every one of these deer was shot and carried away by the Boh-hoys.   I have heard of gentle-

men's gardens being robbed, the thieves presenting rifles
or revolvers at the servants who came to interfere ; and
when I asked if there were no laws to punish outrages of
that description, the reply was, " Yes, there is a law,
but if I proceeded to put it in force (which would take
me a vast deal of trouble to do), fellows would come in
the night and maim my horses or cows ; so I must bow
to that sovereign lord, and worst of all tyrants, the
people."

On Sunday, after the hours of worship, I walked about
the town, but as it was a holy day, though by the lower
classes not in the least regarded as such, I could not see
the machinery, of which I had heard strange reports,
if true, for the killing, dismembering, and packing up,
in a most marvellously short time, of an infinity of pigs.
Not feeling very well, I entered a chemist's shop and
asked for a saline draught, when the man furnished me
with some liquid in a glass, of which, having considerable mistrust, I declined to drink, and then asked him if
he did not know what a saline or effervescing draught
was, and he confessed he knew nothing about it. He
had, however, some effervescing lemonade, so I contented
myself with that febrifuge, and prepared to resume my
journey towards St Louis on the following morning.

CHAPTER VII.

On Monday morning, the 13th of September, I left
Cincinnati by train, attended by a very civil conductor,
and, with my dogs, got very comfortably located in his
car. For some distance the Ohio river continued on our
left, and then fine undulating land, with very good crops
of corn, appeared on either side, the woodlands becoming
more and more thickly interspersed with the settlements
as we advanced. The turkey-buzzards seemed in numbers to increase, or to have been collected here and there
by some carrion in the woods, for they enlivened the air
with their soaring flight, or sat on the dead limbs of the
white bare trees, with their wings slightly outspread, as
cormorants may be seen to do, to dry their feathers after

7

diving. Every mile and every moment seemed to add to the beautiful autumnal tints of the woods, the larger trees everywhere lying in a damp and mossy decay, while the younger growth attempted dwarfishly to conceal the ruins brought about by the reckless hands of the former rovers on the soil—the red men. It is one of the curses of this prolific country, that a tree has no friend, and that all picturesque idea, certainly for a time, must yield beneath the necessity of a clearance for the purposes of cultivation.

The Indians, or those mere savages who remain, the worthless relics of the primeval race who so aptly succeeded to the reptiles of antediluvian worlds, in folly set fire to the plains and woods for the mere purpose of enjoying the passing glare, or of dancing by a fire, doing to themselves an incalculable loss or injury by the wholesale destruction of creatures on whom they live. The white man then steps in, and by lucifer-match, girdling-knife, and hatchet, not from wantonness, but of necessity or in wisdom, adds to temporary devastation, and fells the oak and any forest-tree, that corn may succeed to keep him in tobacco and whiskey, for that at first is the chief aim of the frontier settler. The line of railway on which I was now travelling carried me through the free state of Indiana, and by newly-erected huts at intervals in the woods, till all sign of habitation at last became very rare, and the brilliantly tinted, hushed, and beautiful woods surrounded or seemed to embrace me with their sylvan arms, in the wild loveliness of which I so delighted to revel! Continuously, then, the woods for miles and miles were unbroken save by the straight, trite line of hasty locomotion. With

cither door of the van, in which were myself and my
dogs, set wide open to admit the refreshing air caused
by the rush of carriages—for not a breath of wind elsewhere stirred in that hot sunny day—I gazed from side
to side to catch a glimpse of bird or beast, but with the
exception of the turkey-buzzard or an occasional hawk,
no living wild thing presented itself to notice.

Pigs, which seemed to stray masterless wheresoever they
pleased in an attempt to gain their own living in the
woods, were occasionally on the line, and as we passed
the White River, and more than once stopped directly
over its course or on the branches of its back waters, the
mud of the shoals gave me the pad of the racoon, of the
skunk, I think, and certainly the "seal" of otters. In
all the brooks or streams that I observed throughout the
flat country around this portion of the White River, there
was not even a ripple of clear water; all was thick, sluggish, and muddy, reminding me much of the hue that I
afterwards saw pervading the dangerous tide of the Missouri. Occasionally cows strayed leisurely across the
line before us, or fed on the green herbage afforded by
any little embankment that raised but did not protect
the rail, I suppose the property of some small settler
whose cabin was not in sight. The aspect of the forest
throughout was one of desolate decay, struggling with a
fresh growth of younger timber; and though there were occasional patches of fair wood where the roots and branches
of fallen trees were not so much grown up with those
innumerable weeds and creepers, among them the wild
vine, that make the cover, in a sporting point of view,
so severe, still I at once perceived the terrible difficulty
that must generally exist in regard to the successful work

of hounds, or as to the uses which we expect from them in
England.

I had seen, then, the tracks of wild animals, and also
coveys of the partridge or quail, but these transitory
glimpses or signs of a portion of the sport I hoped soon
to enjoy, were interfered with by others, that made me
question whether I should ever arrive at that sport with
limb and life enough to enjoy it, for cows and pigs were
perpetually on the line, the only place where they
could catch a draught of fresh air or bask in the sun.
Every hour we seemed to be on the eve of collision with
the one or the other. At last I thought it was all over
with us, for in the warm, still afternoon, I beheld *seventeen cows, all grouped together, standing on the rails*, with their
heads different ways, and all apparently asleep. When
I first saw them, in leaning from the door of the van to
look ahead, they might have been about two hundred
yards from us, and I naturally expected that breaks
would have been applied, and everything done to de-
crease our speed; but no such thing. Thinking that
the hour *was* come, and having hastily arrived at the opinion that I might as well see what this sort of concussion
did, or what the "cow-catcher" could effect, as die
without any information on that head at all, I held fast
on to the side of the door of the luggage-van, and continued to lean out as far as I could, so that I had a full
view of what I expected would be a chaos of butcher's
meat, and, to use an American phrase, of "chawed-up"
men, women, and children, while, at the same time, I
thought myself to run as good a chance for escape as I
should have had if I had remained closed up with packages heavy enough to crush me in a confused and *very*
"rough and tumble" fall. All of this, though it takes

long to relate, passed through my mind in a second, and
in that time I knew, to my astonishment, by a sort of
jerk in the train, that the driver increased instead of
diminished the speed at which we were going—or, in
other words, that he absolutely put all the steam on he
could, *and charged the living and phalanxed impediment* that
seemed to threaten us with annihilation! He (the
driver) must have been a downright good specimen of a
cool and high-couraged American, for he referred not to
his whistle until he was close upon the slumbering herd,
and then he let it off with so sudden an impetus that
every cow gave a start of terror, the herd separating
and making a furious effort to escape by jumping on
either side the line. On the right hand side, or the side
from which I was leaning, the last two cows had to
" cringe," or tuck in their tails and haunches, to miss the
" catcher," and that they just succeeded in doing by a
hair's breadth.

It was with a long-drawn sigh of satisfaction, and a
very religious and soul-felt thrill of thanksgiving, that
I then resumed my observation on the surrounding
country, resolving in my own mind that, if I survived
the dangers of the American rail, I should have passed
through the worst of all those impediments which are
said to render the life of man so remarkably uncertain.
When I resumed my place inside the carriages, and
entered into conversation with my kind friends, I told
them that the first thing we did in England was to insure the safety of a railway line by an adequate fence,
and that then, if cattle or men were found on it, the
English law was on the side of the railway company,
and trespassers of whatever sort were punished. In
reply to this the American gentlemen said that neither

their necessities nor their funds would permit of such
precautions for the public safety. The public needed a
speedy transit; and, as the public desired and required
the utmost go-ahead haste to reach a given spot, why,
they got it on the only terms it could be afforded, and
the public must take care of themselves. Lives, compared with mercantile interests, or with the possibility
of reaching the most distant settlements, were not considered. If a dozen men were killed, there were hundreds more ready to fill their places, and hence the power
to "go ahead" and nothing but that power was the
chief object of the United States.

During this journey to St Louis, however, I met with
my first "difficulty," as the Americans would call it,
with a thing called a man in one of the baggage vans.
There were two fellows in this van, who seemed so little
disposed to pay attention to the safety of my dogs that
George Bromfield, the place being crammed with luggage, asked.me again to put myself in evidence, or he
feared some accident might otherwise occur. On this I
left the cars and travelled with my dogs. I had not
been long so stationed ere I had reason to watch the
conduct of the two unmitigated scamps to whose tender
consciences the property of all those travelling by rail
was, for the time being, intrusted. Piled up in heaps
were the trunks and boxes of gentlemen, and perhaps
affluent merchants and tradesmen, and among them the
well-worn and ill-secured little all of some poor emigrant. My trunks and boxes, too, were there, and when
compared with those of the better class of Americans, I
at once became aware that where my packages had one
hoop of iron to protect them, those of my fellow-travel-
lers had ten, with the corners of their trunks rounded

off, so to avoid weak points of collision. As we went on, and approached different places, at which some of the luggage was due, these two scoundrels, in moving and re-arranging their load, never lifted anything that they had the power of throwing down, but threw any trunk or box heavily, and with all the force they could, on its end or corners, and then, as they pulled them into the spot of their destination, keenly eyed every lock and crevice, to see if by their violence they had started either the one or the other sufficiently to give them an insight into the contents, or a chance to abstract and steal property worth having. I had my attention directed on all they did, though, at the same time, and to blind them, I made it appear as if I thought of nothing but caressing my four-footed favourites. There was a little brown and tattered portmanteau, evidently belonging to some poor emigrant, and that had seen its best days, and which a man could very well have moved with one hand; this also they tossed up, and on its coming briskly against a large box ere it fell to the floor, the lock gave way, and the vultures of theft and robbery immediately and closely inspected its contents. I watched them narrowly, and had they taken anything out, my hand would have been ready, at the next place where we stopped, to have led them forth, and to have called on the owner of the little trunk to demand his rights. They took nothing, however, for the whole package consisted, to the grievous disappointment of the intended thieves, of no more than raiment of the most common order.

One of these miserable wretches, the shorter one of the two, was a small man with a large head, lots of hair, and, for his size, a large beard, just as dirty as he was conceited, and therefore, in his attributes, an amalgamation of imperfections. I watched this little wretch with a species of horrible fascination, and perceived that he had an idea that he was fatal to the peace of mind of woman. Whenever a petticoat appeared at the door of a hut, or in the fields or woods, this mistaken apology for a man waved a hand to them, observed in his action by his younger and taller fellow-thief with a species of veneration that made me long to thrash them both; and after making this, as he supposed, graceful acknowledgment that he saw and appreciated the female presence, he invariably took out a little dirty comb, and with it pulled his bushy beard to more uniformity.

While I had my eye on the actions of these most disreputable public functionaries it commenced to grow dark, and when we were approaching the place at which we were to swallow a hasty supper—for, as is usual with all American trains, we were behind time—the tallest thief of the two, lantern in hand, it then being too dark to see distinctly without it, came to me and said, " Guess we'd better settle now for them dogs o' yourn.' " Well," I said, " how much?" " Eight dollars," was the reply. "Stuff!" I cried (I had not had an opportunity of making a previous bargain), " I will give no such extortionate price; you may have six dollars, which is really three times as much as you ought to have; so here is the money, and give me change." " Well," he said, " s'pose 't'll do;" and he held out his hand, into which I intended to put two five-dollar gold pieces, and to receive the change. He would in that case have to return me four dollars. In selecting the money from my purse, however, instead of taking only two five-dollar pieces I had possessed myself of three, the one upon the other, but when I dropped them into his hand they fell apart, and I saw the mistake.

"Stop," I exclaimed, "give me back"—but ere I could finish my sentence he said, "I'll bring you the right change," and he went to a little sort of table he had in one of the corners of the van, and seemed to occupy himself in looking for money.

"That will not do," I said; "I will take no change till you give me back the third five-dollar piece. I accidentally gave you three."

"*You didn't*," he confidently replied; and, with a most vile appeal to Heaven, he took the Deity to witness that I gave him but two gold pieces. I met this impudent assertion at once with a lie direct, adding, "I *will* have my five-dollar piece back, or you shall take the consequences of any attempt at retaining it." On this there was a sort of consultation between the two Boh-hoys, and the scoundrel came back with the two gold pieces in his hand, and said, "There's the only money you gave me; and, rather than have a difficulty, I will take your dogs for nothing."

"No, no," I said; "that will not do. I might agree to that, and save a dollar; for were I to agree to it you would not take my dogs for nothing, because you would still have stolen five dollars. I will have the change out of two five-dollar pieces, and the other gold piece back; or, when we stop, I will drag you by the collar to the carriages, and appeal to the conductor and the passengers collectively to see that I am not robbed." This declaration was evidently distasteful to the scoundrel, for I take it he thought I was fully capable of putting my threat in practice, and he again took the most impious oaths that he had no other money of any sort than the two five-dollar pieces I had given him.

"Well," I said, "I care not for your lies. I will see, when we stop, if I can't get back my money."

My eye was still on the fellow, for I saw he was not comfortable, when he again came back to me with his lantern, and, looking on the floor, asked me to raise my foot, pretending that he thought the gold piece might have accidentally fallen down.

"I will see you at the devil first," I replied, "into whose custody you eventually will be sure to fall, ere I stir a foot. You have got the money, and from you I will have it when the train stops."

We were then slackening speed to approach the platform of the station at which we were to stop for supper, and he advanced to go out (as officials of this sort often do) before the train had ceased its motion, the door being partly open. I then got up, and, laying my hand on the handle, closed the door entirely, saying, "No, my fine fellow, you and I get out together, but not till the train has come to a stand-still." He did not like this, but, seeing no help for it, he again went back to his corner, and, just as we came to a halt, reappeared before me with a lot of silver in his hand, proving how impious were the lies he had told as to his having no money, and which was evidently more than the change out of the two gold pieces, and it was asserted by him to be the amount of the third gold piece. It looked something near it; so, having to get my supper, with a very little time to swallow it in, I told him it would be better for him if I found the change to be all right, when I had time to count it, and then I went into a room, where the hungry passengers were already seated at their suppers.

On counting the money this rascal had given me in presence of an American gentleman, it was found that he had not given me quite the amount of the third gold piece, and my friend wished me again to demand it. I declined, however, in that brief interval to have any further disturbance, but announced my intention, when we arrived at St Louis, to report the theft to the railway authorities, and to demand the fellow's dismissal. We soon recommenced our journey, and, without further let or hindrance, in a cold and rainy morning, between three and four o'clock, we descended from the carriages to a platform of a station immediately on the banks of the Missouri, and opposite to the town of St Louis, and there found several omnibuses in waiting, with four horses in each, that an Englishman might well have driven in his drag in London, and among them an omnibus, for which I had telegraphed from Cincinnati, for the especial conveyance to the town of myself, my servant, and my dogs, my luggage to be delivered in the usual way at the hotel called the Planter's House.

It was beautiful to see these four-in-hand omnibuses drive into the ferry-boat; but I take it that custom and their own natural sagacity kept the handsome teams steady and in the right way, for my spread-eagle friends on the boxes, though they seemed to take a hard-fisted aim at the places to which they desired to attain, seemed to have as little analogy to that artist called "a good coachman" as a costermonger behind a donkey has, whose progression ceases when he lets go the thing he calls a rein attached to the head of his enduring ass.

Having crossed the river, into the hotel called the

Planter's House at last I walked, when the insufficiently attentive man at the bar told me that through the telegraph I had been expected, but they had not deemed it expedient to have a place ready for my dogs. It was then four o'clock on that most uncomfortable morning, when, by dint of ex-postulations and of a bribe to a porter, I commenced my search in the streets of St Louis for a sufficient place in which to put my man and dogs. We went to two tene-ments in vain, but at last were directed into one of those wretched places for horses and mules which the Ameri-cans deem it no sin to call stables, and in that dirty and confined elongated hovel I discovered a loose box, and to its shelter consigned George and his charge, while the "watchman," as he was called, on the premises, went to fetch him some straw, and myself to obtain him some bread and coffee. " You may be quite easy," said the man at the bar of the hotel, " that the watchman will take care of your dogs, and your man can come in for refresh-ment." " Thank you," I replied, " my man will *not* quit his charge, and to-morrow he will seek better accom-modation; I will see that he has his coffee where he is."

Soon after my servant and my dogs had been left to their repose, and I had gone to the Planter's House, in spite of the boasted watchman, George heard a footfall stealthily approaching in the dark, and then some one clambered up on to the top of the loose box.

" Who's there ? " exclaimed George.

" Oh, are you there, young man ? " replied his un-known visitor; " if you'll come with me we'll have some-thing to drink."

" I've had plenty to drink," said George, " and what's more, if you come down here I'll lay my stick over your head."

"Oh!" answered the intended dog-stealer, sliding down again on the side whence he came, and George heard of him no more. When this was reported to me the next day, I almost regretted that George had not lain quiet and permitted this scoundrel to have let himself down, if he would have done so, to the tender mercies of Druid, Brutus, and Bar, who, on finding strange tobaccoey legs suddenly thrust into their straw, would very soon have made very short work of this son of freedom, and taught him the wholesome necessity that there was for no more of his attempts to steal the property of a stranger. No doubt if he could have decoyed George away, an effort would have been made to take some of my dogs, but how, in the short time before daylight that remained, they would have been able to blind or appease the wise ferocity of Druid and Brutus, I have not the least idea. I would have given worlds to have been hidden close by, to have watched the dangerous endeavour.

On the following morning I made one of the pleasantest acquaintances of those very many agreeable ones that I formed during my stay in St Louis, and saw Mr Robert Campbell. He at once set to rest all my anxiety as to my dogs by giving up to their service the large rooms of one of his houses or stores at the bottom of the town, looking pleasantly on the Missouri river, from the ample window of which George could amuse the succeeding hours of his watch by looking at the river, shipping, and constant bustle of that busy and crowded wharf, while at the same time the windows, not being on a level with the floor, could be set open, with a slight additional protection, for the purposes of healthful and complete ventilation. The precautions in the disposal of his dogs taken by George were excellent. Druid and Brutus were al-

ways chained on either side the door, and George's own
bed was made on the floor, about the middle of one side
of the room. Bar, Chance, and Alice were tied up in different places, so that there could be no entanglement of
chains, nor interference between one dog and the other.

On the first morning of my paying a visit to my dogs,
after they had taken possession of their ample rooms, I
shall not easily forget the appearance from the attacks of
mosquitoes of my servant's face and hands. It seemed as
if all the mosquitoes around St Louis, attracted by lodgings
agreeable to themselves so near their favourite site (a
river), had congregated there with very little to subsist
on until the arrival of a man and dogs. When that
amount of life and warm blood was infused into their
hitherto provisionless haunts, then, indeed, they revelled;
for in summing up his grievances to me, George said that
the mosquitoes so attacked the dogs that they did little
else than scratch, shake, and snap throughout the entire
night. In these upper rooms, belonging to Mr Campbell,
no attempts were made on Druid's privacy similar to an
accidental one which I forgot to narrate in its right place,
as having happened in Mr Easton's room at Cincinnati.
At Cincinnati, the foreman, I think, or clerk, to Mr Easton, unaware that the room was so tenanted, without ceremony flung open the door, when his advancing leg,
amidst a roar from all the dogs, was within a hair's
breadth of being seized by Druid and Brutus. Had he
not been immensely on his haunches, and taken the brisk-
est skip to the rear, the consequences might have been
exceedingly unpleasant to me, as well as to himself, for I
should have been the cause, though indirectly, of injury
to a trustworthy and good man.

On my arrival, and through Mr R. Campbell, I reported

the thievish propensity and dishonest conduct of the fellow in charge of the baggage, when Mr Campbell informed me that in addition to my complaint there had been another lodged against the same delinquent, and that he had little doubt but that he would be dismissed from further employment.

On reaching St Louis, then, I had arrived at the last large town, on that route to the plains, of refinement and civilization, and I found that some waggish citizens of the United States rather contemptuously called it "the turning-back place of the English sportsman." The eyes of the American public, in fact, were fixed on me, attention having been particularly aroused and drawn to my position by my letter to the sportsmen of the New World, which had gone the round of the local press, and by publications of my intended visit which had appeared in the London newspapers. I in no way disliked the notice that had thus been accorded, or the doubts as to my further progression, which, though not openly divulged, I could see were nevertheless by some entertained; it afforded me much quiet amusement, and I had considerable internal satisfaction in drawing out some of my transatlantic friends into an inflamed and vivid picture of the dangers I should have to encounter. The hostile attitude that two thousand Indians in war-paint had assumed, their numbers probably overstated, directly in front of the line by which the hunting grounds could be attained, was also much dwelt on. Searching glances of some of these acquaintances of mine, which followed their vivid narrations, were intensely satisfactory to me, but I do not think they afforded any very great information to their proprietors.

An amusing story also was told me, of a young, a *very young* gentleman, who had made his way from England to St Louis on an alleged hunting expedition to the plains. He obtained an outfit at St Louis, and was proceeding by river steamboat to the desert of the red man and bison, when of course he had to mingle in the society around him in the saloon, which is generally of a *very* mixed character, and some of his apparent friends neither lost their opportunity nor mistook their man. Having been, as they say in those parts, "pretty considerable down on him" as to the hardships and horrors that would attend his journey when on the prairies, one of his attentive friends, on finding that a " con-si-de-ra-ble " impression had been achieved on the mind of the English traveller, at once became seized with a roving as well as an amiable desire to rescue him from danger and difficulty by taking his outfit then and there off his hands, and going to the plains in his stead. This kind offer having been embraced by the *very young man,* an agreement was at once entered into between him and this gentleman—a passenger, of course by accident, on board the steamer, much addicted to cards and dice, as well as deeply interested in the fair management of a lottery, at which the greatest bargains were to be won on very small entries.

A deed of sale, but, alas ! a very one-sided one, having been on the spot arranged, the casual acquaintance and very kind friend of this young man at once gave him a cheque on the firm of Springfield and Co., St Louis, when, having surrendered his outfit into the hands of the amiable stranger, the young man, on the first favourable opportunity, reversed his line of progression, and returned to the town he had left, wherein resided the copartners who were to honour the bill of exchange of which he had become possessed. Reader! dear reader, perhaps you are prepared for the result? There was not such a firm in existence as that of Springfield and Co., and the dupe thus retrogressed to the " turning-back place of the English sportsmen," I hope a wiser, but certainly a poorer man than he was when he left it.

Having listened to everything, and believed just as much as I deemed necessary, I then considered the advice I had received in the old country, and ascertained that the idea of going out with " trappers " after bison or buffalo was nonsensical as well as impossible at that season of the year; and that to attach oneself to any tribes of Indians, Mr Murray's friends, the Pawnees, *par excellence*, was as much beside the mark, and as imperfect in regard to personal safety and the safety of horses and mules, as the reception of a cheque on Springfield and Co. could have been with a view to its money payment. The only fact that gave me serious or painful consideration was, that the distance and cost of my mission had been immensely underrated, and I really had no more funds at my disposal than would have taken me home, and, indeed, hardly those; what then was I to do? The question was at once answered, and I resolved that it was neither consonant with my own character, nor according to the custom of my past life, to turn back from any adventure on which I had set out. I went on; but how I achieved progression must remain for narration in the next chapter.

CHAPTER VIII.

MINE, as the Americans would say, was far from a pleasing "fix," to find that the cost of my expedition had been immensely underrated; but again, in the course of my travels both in France and America, I had to admire the offhand, unselfish, and high-toned liberality of American gentlemen. In a former work, entitled "A Month in the Forests of France," I had pleasure in acknowledging this fact in the instance of the captain of one of the sailing packets of the United States, whose address I have forgotten; but now, in the same liberal spirit, though to a far greater extent, it is with infinite satisfaction that I allude to the kindness of Mr Robert Campbell, of St Louis, who, apparently guessing what my wishes would be, rather than having them explained to him, at once offered to place at my service any sum of money I deemed requisite to enable me to reach the plains, and to fulfil the intentions which had brought me across the wild waves of the Atlantic. In the course of the chequered phases of my life, I scarce know one of the friendly acts of my acquaintances (and from them I have received many) which filled me with so much delight as this splendid offer of Mr Robert Campbell's, and I hesitated not at once to embrace it.

Of course the thought had heavily weighed on my mind as to what envious detractors in England would have said, if I had turned back from any difficulty or danger, and I know but too well what my own feelings would have been in the event of being forced to miss my longed-for opportunity of hunting up the giant game on the prairies of their own desert. Those who have shared in my first hare-hunt, sported with me in my boyish hours, and ridden in manhood by my side when hunting stag or fox, they, at least, would understand my jump from sadness into joy, when I thus, almost unexpectedly, saw all that I could then and there desire placed subservient to my will; the means most amply put at my disposal by which I should slay the wild bison, and all this by a friend of but a few hours' personal knowledge of me, although, being an enlightened gentleman, he of course understood my station in society. Mr Campbell then introduced me to many of his friends and acquaintances, some of whom had been to the plains, as well as to his brother, who also received me with a kindness I shall never forget. We then consulted, with the strictest view to economy and efficiency, how and in what way I should collect an outfit, and most speedily and cheaply travel to the furthest limits of the settled towns, Kansas city, in the territory of that name, being the civilised spot whence I was to "jump" into the desert.

It was soon resolved that the first thing to be done was to procure an ambulance waggon, to be drawn by two good mules, in which I was to carry my light baggage and to sleep; this ambulance to be fitted with a double cover or tilt, and to travel on easy springs. The other waggon was to be of stronger construction, to be drawn by four mules, and to carry the heavy baggage, consisting of tent for my men, provisions for six weeks, and all the necessary appendages for camping out on the Plains; to these two waggons had also to be appended the necessary harness for four and a pair of mules, to which I subsequently added a single set of harness for one stout mule, which was to draw a smaller waggon for my dogs. Picket-pins and ropes, halters, camp-kettles, water-buckets, axes, hatchets, spades, bacon, flour, sugar, salt, pepper, and a hundred other things, had to be collected and arranged.

I also had to procure for my own immediate use two saddles, made on the English principle, and a double rein or curb and snaffle bridle, and, much to the astonishment of the best tradesmen in St Louis, had an opportunity of ridiculing the American saddle, which is at least a hundred years behind the modern or English improvement, and of laughing at the huge upstanding crutch, which rises perpendicularly from the pommel, of course for the purpose of holding on by in an unseating difficulty. This crutch insures as well the great probability of the death of the rider, in the event of his horse throwing a somersault and falling on him. No unlikely thing I thought to happen to horse and rider on coming into the holes of a prairie-dog town, partially concealed by long grass. Of all the unsightly, hideous, and dangerous things on a saddle, this excrescence is the worst. It is as large as one of the crutches to a lady's side-saddle, only straight, and must be fatal to the life of a man in such a fall as I have referred to. The American saddle is also high behind; it has flaps like those of the English make, but much smaller, and when I got into one I felt just as if I sat on a deal board, with the bowsprit of a ship ready to rip up the buttons of my waistcoat, or penetrate my waist to the impossible arrangement of any future dinners.

It is evident to me, by the make of their saddle-flaps and the length of the stirrups in which they ride, trying in an ungraceful way but just to touch the coalbox-looking or clog-like thing called a stirrup at the end of the leathers, that most of the horsemen in the United States *wish* to hold on by the calves of the legs (or where their calves ought to be), and that they take no grasp with the knee and thigh whatever—the pommel crutch being regarded by them as the hold to be relied on in a difficulty. Though I searched the shops in St Louis, I could find no copy of an English saddle with spring bars, such as we have in our hunting saddles. That safe, improved, and handy invention for quickly slipping out the stirrups for cleaning has never been adopted in the United States. In saddlery and the art of riding, in those regions visited by me, they are, as I said before, a hundred years behind us. They have but one class of single-bit generally in use; they tie down the heads of their horses on all occasions (whether the carriage of the head needs it or not), with martingales, and always strangle them to a considerable extent by buckling up the throat-lash to the very highest and tightest hole to which it can be made to attain. As to driving, somehow or other they guide their horses (though, to an English eye, in a spread-eagle-like and most unseemly way) safely enough to where they wish to go, and on many occasions dispense with winkers.

Their horses and mules are generally very docile; and on one occasion, in St Louis, and in the possession of my friend Dr Pope, there was a favourite horse of his that could be left in the street with his gig or cab without any attendant, and that would, when he saw his master coming, turn himself off at an angle to open out the step for his more easy ascent into the carriage. He would also take his carriage from the house-door to his stable on being told to do so. I observed in the streets, in all the towns through which my journey lay, that the horses of America were most sensible and obedient to the voice of their drivers, though, with the exception of some gentlemen who spoke kindly, the words addressed to them by the lower orders were of the coarsest and most brutal description. The sort of horse thus brought within my observation, for both riding and driving, could scarce be surpassed in England. There is, though, one fact on which I must congratulate the Boh-hoys in New York; it is this, that if the same number of all sorts of vehicles were jammed up together, as I have seen them, in the most dray-frequented part of the Broadway, in any portion of the city of London, there would be more swearing and slang among the English drivers of that description than in those of the United States.

It is the fashion in America to boast that the presence of woman ever meets with respect; of course her presence always commands it among gentlemen of any country in the world, but in New York a policeman is obliged to attend at each crossing of the Broadway to prevent women being run over by the Boh-hoys. The language uttered in the presence of ladies in railway carriages by this influential class of society is utterly detestable to the ears of an Englishman.

In addition to my ambulance waggon previously alluded to, I invented and had made to order an under sort of tent or wall, to fasten to the sides and wheels of my carriage, to protect my dogs at night or during a halt from the weather, the bottom of my ambulance of course being the roof over my dogs, thus comfortably inclosed. They, those " dear companions of my leisure hours," well and truly repaid the protection and care I gave them, for, chained at night to each wheel of my ambulance, Druid, Brutus, Bar, and Chance, not only would let no one approach me, but if any unusual noise arose in or around my little camp, they were sure to awaken my attention. To their vigilance, love, and fidelity I am sure I owe the safety of my horses and mules, and perhaps my life, as the sequel will show. While the orders to tradesmen which I thus gave were in execution, I had leisure to acquiesce in Mr Campbell's kind proposition to take occasional drives in his carriage around and about the city of St Louis, and to see all that the shortness of the period for which I designed to remain would permit.

I was awaiting his call for me at the Planter's House one day after dinner, when a man with a paper-bag in his hand was announced as wishing to speak to me. He came, and I saw before me a shortish man, whose kind, civil, and honest look pleased me much. He came up and said, " Georum wan mo scrabs for'm dogues." " *What?* " I said, and he repeated the intended intelli-

gence, but though I laughed and he smiled I could not
for the life of me make out what he wanted. At last, by
slowly repeating the word "dogue," pointing at the same
time in one direction and holding up the bag, which I
saw had grease upon it, I made out that "George wanted
some more scraps for the dogs," so I took my good and
attentive friend to the bar and obtained a further supply.
John was a German, and I certainly never met with
a more attentive and obliging servant. Humble as his
calling may be, John will ever be remembered with the
greatest kindness and good-will. What a strong contrast
he afforded to the waiting Boh-hoys raised in the United
States of America, some of whom I encountered, and, to
their astonishment, briskly admonished, during my visits
to St Louis, when, for the time being, they were induced
into more civil behaviour. On several occasions, some
of which I will relate in due time and place, I had to re-
prehend these fellows with instantaneous effect; and,
perhaps, if the American gentlemen would be more particular in exacting attention, and bear themselves less on
the equality system, dinners would be better waited on,
and the serving men no worse. During my drives around
St Louis and its vicinity with Mr Campbell, I had the
pleasure of making the acquaintance of Mr Shaw, who,
with a beneficent and public spirit not often to be met
with, is at this time arranging his Missouri Botanical
Gardens, extending over 800 acres of excellent land, to
be dedicated to floriculture and public recreation, and I
believe eventually assigned as a bequest to the inhabitants of St Louis. I paid more than one visit to Mr Shaw,
and had the happiness of walking in his gardens with a
fair inhabitant of the United States, who added to her
other attractions a mind full of information, and gave me

a most favourable impression of the female society of her country, and what a boon Mr Shaw was adding to his splendid public gift by affording to future wanderers a quiet scene for converse and contemplation.

While at St Louis, Mr Campbell also took me to see the large public rooms and mercantile library, built by private subscription, where in and among other choice selections of art I saw a statue of the alleged parricide Beatrice de Cenci, representing her the night previous to her execution for the murder of her father. Beatrice is represented with the cross in her hand, and is a graceful production, executed in Rome by Miss Harriet Hosmer. There was also a statue of the nymph Œnone, of Mount Ida, the size of life; and the artist has represented the nymph at the moment of her desertion by Paris. The attitude of Œnone and her form, to my mind, are perfection; the latter chiselled in the fullest mould of womanly beauty. The hand, however, as well as the foot, arrests the eye of the observer, and by their obtrusive claim to notice they give an indication of being too large. The wrist, too, though the arm is as round and full as it could and ought to be, is not quite in accordance with the delicate symmetry of the rest of the figure. There is also a statue of Webster. It was in the large room of this handsome public institution, and while I was inspecting the building, that my friend Mr Campbell suggested to me the idea, afterwards carried out, of my giving a public lecture.

St Louis struck me as one of the most rising cities that I saw during my journey through the United States; and when its public buildings, as well as the numerous private houses which are in course of erection, are considered, it is very evident to me that it will one day become—perhaps it is so now—the queen of the Missouri river. I believe it has been only twenty-five years a city, and, when viewing its dimensions and the public desire for improvement, and the patriotic liberality of such gentlemen as Mr Shaw, to whom I have before referred, with the numerous and splendid river steamers trading to its wharves,—why, if St Louis does not become one of the gems of the Western World, it can only be from the fault of the sons of the soil on which it stands.

I observed in my drives and walks through this growing city, and also derived the intelligence from my personal friends, that the Roman Catholics, with some of whom I have since become acquainted, were not only possessed of very great property there, but that, as a body, they were very considerable and influential.

During this my first stay in St Louis I also made the acquaintance of Col. Sumner and Capt. Clery of the United States army, and of other officers, and from them, as well as their brothers in arms at New York, I received letters of introduction to the officers of all the forts on the desert, to insure me not only the kind reception so natural to give and to be received by every soldier and gentleman of the same high service, in whatsoever country they may meet, but to afford me places of protection in the event of my meeting with any unforeseen discomfiture. It was in course of conversation with that able and gallant soldier, Col. Sumner, that he gave me the intelligence that no man had ever yet achieved the death of eleven buffaloes or bisons at one run at a herd from the back of one horse. Here then was a wicket set up for me to bowl at, and often at night, when in bed, but not asleep, I contemplated the possibility of achieving that number. I was, however, too old a sportsman **not** to be fully aware that when any-thing was to be accomplished by man and horse, a great deal, if not all, would depend on the rider and the ridden being of the same mind. Thinking it well over, the only conclusion to which I could come was, that, as I " could not command success, I would do more, Horatio, I would deserve it." To my readers— those of the strong and graceful hand and the high and chivalrous heart, with the world yet before them—I would recommend the study of the above quotation both in love, the chace, and war, for in failure, perhaps, the assertion thus made to Horatio is the only balm to a wounded conscience.

Every day of my stay in St Louis either added to my list of kind friends, or gave me fresh opportunities of remarking on the peculiarities of that class of the nation devoted to trade, and to observe on the impudence of some of the positions taken up by men in the pursuit of dollars under difficulties. There is in St Louis a remark-able instance of this; it is by no means an isolated one, but it will do for a sample of the others. A highly approved gunmaker establishes his shop, and gains con-siderable custom ; another of the same trade, not having been so fortunate, observes that the adjoining shop to the successful tradesman is to be let, and he immediately establishes himself as tenant, the entrance-door to either shop being side by side, and the contents in the window of the new-comer made precisely to match those in the window of his neighbour. In short, the two windows appear to belong to one establishment, the name of the old inhabitant being purposely left in superior characters to that of his aspiring friend. Customers, if they ap-proach these Siamese but schismatic tradesmen on the side of the crafty aspirant, are sure to fall into his hands, and I must in fairness own that on two separate occasions it was my fate to do so.

After a great many consultations in Mr Campbell's counting-house, with several persons supposed to be conversant with prairie life, I soon began to see that veracity was very far from being an object of veneration, and that if any man had been as far across the desert as the Rocky Mountains, it was impossible to believe a word he said. Either the climate of those wild hills, or the fact that when a man quitted the frontier settlements he left all observers behind, so inflamed or enlarged his imagination that the action of the hour was forgotten in the mists of mental reflection, and the traveller became a visionary instead of a verity to be relied on. From one of these gentlemen I had been in the habit of taking advice, when all that he had told me was considerably shaken by the following story he was very fond of telling.

He was out on the plains in winter time, and by some accident severed from his party; night came on, and with it the most intense frost, and he found his vital energies failing, and saw nothing, until a bison came to his assistance, but destruction staring him in the face. The friendly bison approached without any idea in his shaggy head half so extensive or comfortable as pertained to the uses to which he was presently to be put; and my adviser, with just strength enough left in him to level his rifle, numbed fingers and all, was lucky enough to hit the bison just through the heart, and drop him dead conveniently for future operation. Having refreshed himself with a bit of warm raw flesh, my adviser then cut the bison open, and took out the entrails; and as it was

just getting dark, to save his life from the intense frost,
he replaced the entrails with his own body, and crept
into the carcass of the animal with his head towards the
tail, and his feet pushed up his windpipe, the skin, when
it had been divided, being permitted to fall as a curtain to
the ground. Here he revelled in animal heat, not in the
least incommoded by that enormous quantity of blood
and moisture which I have since seen contained in the
carcass of an animal of this size, and at last fell asleep.
How long he continued in comfortable oblivion he is not
quite sure, always desirous of being exact, but he awoke
with a sensation of cold, and found that every limb save
his hands, which he had fortunately kept folded under
his chin, was not only benumbed, but the blood and
moisture had become a frame of ice, and the severed
skin of the animal itself had frozen firmly and immovably to the ground. Thus, then, he was iced for the rest
of his life—a sort of forced meat, in fact, in the bovine
"rissole" which surrounded him, with no white-capped
and aproned cook to deliver the buffalo of his despairing
burthen. Simultaneously with his awakening from his
sleep, however, he had been aware of a strange noise
outside the frozen case which compassed him; so, with
ears sharpened by necessity, he endeavoured to discover
the cause. The external rattling and scratching continued, till slight openings admitted a gleam of the intensely cold morning moon, when through the crevices
he could plainly distinguish the gnawing, crunching of
bones, snarling, and fighting of a large drove of wolves.
Here, then, was a Scylla to the Charybdis which already
contained him, and a more active or pending dissolution
was thus fearfully around; for, sportsman as he was, he
well knew that there was always an immense and hasty

scramble among animals of prey for the choice intestines
of a body. Bold, not with brandy, but whiskey (for he
at once and on the moment emptied his flask), and desperate with distress, he resolved on action, and abided
time; and here again, in this his dreadful hour of need,
his knowledge of sport and the habits of wolves did not
desert him. He recollected that the easiest portion of
a bison's frame, by which for a wolf to gain a hold of the
intestines, was from beneath the tail; so he awaited the
result, and kept his hands in readiness. All at once,
with eagerness only parallel to his hunger, an immense
old master wolf rent the flanks asunder sufficiently to
thrust his head and shoulders into the bison, in search
of, to him, a delicious reward. The wolf's horror, however, at once may be conceived when my friend as suddenly seized him on either cheek, and clung to the head
of his deliverer for dear life. So surprised at this unwonted reception was the beast, that he yelled with terror, while, with all the strength of his back and
haunches, aided by his fore feet, he contrived to withdraw, pulling the human appendage along with him. The
result of the yells of the master wolf, as well as his great
strength, were alike favourable to my friend; the former
so frightened the rest of the wolves that they fled away,
and the latter brought my friend forth from his "lodging
on the cold ground," and rendered him up to the grey
dawn of the coming morn. And what became of the
delivering wolf? "I let him go, sir. Yes, sir, as soon
as he pulled me out of the carcass of the buffalo, and he
fled after his scared companions. Yes, sir, that's a
fact!" I recount this "tale as it was told to me;"
and, in justice to my friend, I must say that I fully
believe that he has told the story himself so often that

he is convinced of the truth of it; that, however, is nothing to me, further than it, at the time, raised a fear that the rest of the intelligence, propounded for my future guidance, might have been of the same substance as the story I have just narrated.

My waggons, harness, provisions, and camp-fittings having been by this time prepared, as well as a considerable box which I was erroneously advised to take by way of presents to the Indians, it then became necessary for me to consider the best way of reaching Kansas city, the last considerable town on the frontiers of the desert. The matter to be considered was—Should I take my baggage by railroad or river steamer? The latter would consume the longest time, and extend over 400 miles of conveyance by water. Having had a surfeit of the horrors of the rail, and being at the same time desirous of ascertaining if the ideas in England were correct in regard to the dangers and reckless go-ahead desperation of the captains in command of river steamers, I resolved to make the method of my further journey dependent on the kind of accommodation I could obtain for my dogs. My waggons and heavy baggage would go infinitely cheaper by river transit than by rail, and if I could make it convenient I resolved to go by the same conveyance. Mr Campbell soon settled the matter by saying that he knew the captain of the "Skylark," Capt. Sousley, and would obtain his kind attention to all I desired. A kennel having been fitted up for my dogs, and a double berth to myself having been secured, I resolved by personal inspection to judge whether or not the dangers and riotous conduct of the river steamers in the United States were as rife as my countrymen represented them to be. A four days' voyage over 450 miles of the Missouri, I thought, would give me a pretty fair insight into river navigation.

Having given orders for everything to be sent down to the wharf for embarkation, and, with the promise of a bribe, desired the head porter at the Planter's House to take charge of all my trunks and boxes, and to see them put on board the right steamer, I prepared to take leave of my kind friends in St Louis, to work my way by water to the verge of civilization by reaching Kansas city, in the territory of that name, and there to purchase such horses and mules as I might require, and by so doing avoid, as much as possible, the cost of carriage. Having, as I supposed, taken care that no mistake could arise in the shipping of everything on board the "Skylark," I amused myself, till the hour of departure approached, in walking about the town and bidding people good-bye. Just on the eve of the departure, then, I reached the wharf, and found, to my indignation, that the Boh-hoy of a porter from the Planter's House had not only left behind a fishing-rod and rifle, but he had absolutely put all my things, except the waggons, on board a steamer going the very reverse of my way, and that that steamer's steam was up, and her wheels absolutely in motion. All action on my part was at that instant anticipated by Mr Campbell's faithful German servant, John, who, on discovering the mistake, leaped on board the vessel, and calling all bystanders to assist him, resolutely returned everything to shore, and brought it away in triumph to the "Skylark," while the head porter of the Planter's House, to whom I had promised a douceur, skulked away without asking for it, apprehensive, most probably, of a very unpleasant reception had he at that moment crossed my path. This last act made me resolve never to go to the Planter's

House again, for the attention shown by the men behind the bar, as well as by the waiters and porters there, was very indifferent; so I recommend all future travellers to St Louis to put up at Barnum's Hotel, for that is infinitely the better house, and there they can be as happy and as comfortable as at any hotel in the United States. I remained there some days on my way home. Having been introduced to Capt. Sousley by Mr Campbell, and made up my mind that I would sooner drink a glass of wine with him than fight him, for he was a tall, long-armed, lathy son of the States, I saw my dogs in their boarded-up kennel, and my waggons go through the extraordinary process of being craned from the lower to the upper deck, and then had the amusement of seeing a fisherman bring alongside a catfish of some fifty pounds' weight, which he had just taken on a line, and a nastier-coloured, bluish-looking, uglier fish I thought I had never seen. He indeed must have been a bold or a very hungry man who first made the experiment of mastication on the slimy body of such an ill-looking inhabitant of rivers. Mr Campbell took leave of me on board the "Skylark," having, in company with the captain, selected my berth; and shortly after orders were given to let go, and on a beautiful afternoon on the 16th or 17th of September (I forget which), and in great good spirits, I commenced my first voyage on the dangerous bosom of the Missouri.

Before proceeding further, my readers, as well in England as in America, will, I am sure, pardon me for a short discursion thus made from the regular course of my narrative, in order that I may show that my opinion as to some impending danger to the union of the States was not unfounded, and that my remarks, which appeared in the commencement of my work, have been borne out by recent and concurrent circumstances. The *Missouri Republican*, of St Louis, bearing date the 27th of January last, gives an account, copied from the *Louisville Journal*, of a grand banquet and "fraternization of the executive and legislative powers of the great States of Kentucky and Tennessee," at which were also present "the State officers of Indiana and some of the municipal dignitaries of Nashville, Cincinnati, and a delegation from the Ohio Legislature."

In the invocation of a blessing and prayer pronounced by the Rev. Mr Royt are the following words:—" We come before Thee in an intercession *hitherto unknown*. Save us from every dissension, from civil war and from blood; restore peace to our borders, and bind together this vast population in the bonds of fraternal and untiring affection."

This allusion to disunion was followed up in a speech by Judge Bullock, in which were the following words: " The States of Tennessee and Kentucky are now drawn still more closely together by the apprehension of a common danger."

Judge Bullock concluded by reading the following " sentiment : "

" *Tennessee.*—In this *national crisis* she will cherish in her heart of hearts the noble sentiment of her patriot hero, *The Union, it must be preserved.*"

That sentiment was followed by several others, whence I need select but one more :

" *Kentucky.*—If treason to the Union shall prevail in the North or in the South, our noble State will stand between the two sections, as stood the people of old between the living and the dead, to stay the progress of the pestilence."

Lieutenant-Governor Porter also admits the presence of danger in the following words: "In times like these, when even the stoutest-hearted and most hopeful statesmen and calmest politicians seem to apprehend danger, and look with dread upon the consequences that may flow from the existing agitation and condition of affairs—consequences, if come they must, which will leave in their track desolation and blood."

My readers will, I hope, then, deem that my fears as to the present aspect of American affairs were justified at the time at which I wrote the previous chapters; and here let me ask, *Where is the root of the evil* which now threatens to shake the vast continent of America to her very centre? The reply to this is, *Her too great democracy, her republicanism,* and *the power which has been assigned by the universal suffrage to the people;* to her exclusively democratic returns to Congress, and to the domineering will of unthinking multitudes, and their intimidating dictation over the conscientious use of the franchise.

I cannot agree with Governor Willard when he asks, "Are we not of one race, of one blood, of one family, of one destiny?" My reply is, "*No.*" It is one of those facts that is fraught with so much danger to America, that an immense portion of her population is of adventurers, of foreigners, and of the refuse or surplus population of other countries; and that to these adventurers, and men with nothing to lose, in too short a time is intrusted a vote in her political fate. All this arises from one great error—an overwhelming democracy. Her people—those that ought to be her labour-

ing classes—are " drunk with liberty and unrestraint ; "
injuriously set up above their true position, they are
inclined tyrannically to dictate to and interfere with the
possession of property, in the wealth of which they do
not share.   I learnt all this not from hearsay—I held no
conversation on political affairs with any soul while I
was in the United States.   The opinion so recently
expressed by me was founded exclusively on personal
observation.   It is shown that American politicians
deem that a crisis is impending ; may Heaven grant
that my apprehensions be not realized to the full extent,
and that there yet may be no hand madly rash enough
to " cut the withe and sever the bundle of sticks."   That
is a homely phrase, but having used it in regard to the
present position of the United States to the large and
kind audience who attended the lecture I gave at St
Louis, I use it again, and again assert my hope that the
danger may pass by.

On inspecting the river steamer, the "Skylark," in
which, on Saturday, 17th of September, I commenced my
voyage from St Louis to Kansas city, I found an ample
saloon or dining-room, extending the entire length of
the vessel, beautifully fitted up, with a stove at either
end, the after-part of the saloon dedicated to the female
passengers.   The berths, which had two doors, opened
into the saloon on one side, and on to the lower deck, or
open gallery, on the other, and were double, but I took
the precaution to secure mine entirely to myself, the key
being always in my pocket, cautioned as I had been to
keep it there, as there might at any time be thieves on
board.   Over the saloon was the upper deck, to which
my waggons had been craned in company with a waggon belonging to a tall emigrant, which contained his

little all, between whom and myself, and while the waggons were being lifted, there had commenced a sudden
friendship, originated, perhaps, from my having more
attention paid me than he had, from our having a common interest in the waggons, or from my taking rather
an active part among the crew in sharply reprehending
any unnecessarily rough usage. After the location of
our waggons on the upper deck, and during the voyage,
he often lamented that we were not to take the plains at
the same place, and travel together for mutual safety
and pleasure. Poor fellow, he had reason to be more
thoughtful than I had, for he had a wife and two children, as well as all he possessed in the world, to take
care of, and we were about to cross the prairies in a
troublous time. May Heaven grant that he has come to
no harm! This man was a true specimen of the frontier
settler, in mind, address, and figure. Tall, taller than I
am, high-shouldered, and cadaverous-looking, the large
bony frame of a man with the muscular portions fallen
away, or wanting that roundness which makes the
human figure beautiful. In manner frank, mild, good-
humoured, and obliging, with a lively appreciation of
anything manly he observed in another, and, I am sure,
a readiness to stand by his friend in any amount of
danger. To this he added a virtue not generally
remarkable in men of his class in America, and that
was, that from the first he *put on* no familiar ostentatious
assumption of equality. He was respectful, but at the
same time manly and straightforward, and it seemed as
if we each were known to the other in a moment.
What an immense benefit numbers of the New York
tradesmen, as well as hotel-keepers generally throughout
the United States, would have obtained from the study

of the bearing of this emigrant. In him they would
have seen the facts illustrated, that incivility to gentlemen and vulgar assumption fail to make the impression
they intend; whereas an attentive, mild, unassuming,
but bold and manly manner, places a man, however
poor, as the fit associate of recognized rank throughout
the world.

CHAPTER IX.

AFTER steaming for twenty miles up the comparatively
bright waters of the Mississippi, we arrived at its junction
with the Missouri, and there I had an opportunity of
observing the different colours of the two rivers, and
how reluctant the waters were, when running in the
same direction between the same banks, to mingle the
one with the other; their antipathy to each other
reminding me of the antagonistic properties of oil and

vinegar. For many miles after they have joined the
waters of the two streams can be distinguished the one
from the other. Now, then, as we stemmed the direct
path on the Missouri river to the bison's domain, I felt
myself prone to ponder over some of the stories I had
been told of them, even though we were passing the
spot where St Louis and Capt. Clarke wintered during
the first exploring party, and the cave that had sheltered
them. The banks just at this portion of the voyage
were beautifully wooded, with bold high rocks to the
water's edge, breaking from the surrounding foliage, and
rendering the scene enchanting. On the buffalo or
bison, however, my thoughts chiefly ran, and I mentally
dwelt on a tale that had been told me, namely, that one
hundred thousand buffaloes were computed as the numbers annually slain by man! Let my readers conceive a
herd of these huge ferocious-looking animals reckoned
at ten thousand strong—I have been told that at times
they have been seen so collected—and then a panic seizing
the leaders of the throng, always those most prone to
fear, cows, heifers, and calves; imagine their flight *en
masse*, or " stampede," charging blindly on anything in
their way, the rear propelling the front, and of course
only the first few aware of any obstacle to their headlong course. When a " stampede " to this amount
directs its charge on the line of march of the United
States' army in crossing the plains, the great guns have
been obliged to open on the head of the advancing
column to turn them from their way; for, when thus
rushing madly on, had there been a collision, baggage-
waggons, guns, and any living or dead thing that came
in their way, would have been trampled to atoms.
What sport it would have been, I thought, to have

ranged at speed on a good active horse up to the side of such a furious and phalanxed herd as this, and to have used my handy little breech-loading carbine; it was with difficulty that I wrenched my mind from such reflections to the scene of the passing hour.

The landscape on the Missouri river, in the districts through which I passed, was generally flat and monotonous, for the most part through woods and between low banks of unvarying aspect. The Missouri is navigable from St Louis for 2300 miles, or as far as the mouth of the Yellow Stone river, which again for 250 miles further is navigable up to the gorges of the Rocky Mountains. The "Big Flat River," the "Little and Big Sioux river," the "Big Cannon Ball," and the "Little and Big Missouri," all join the Missouri, and render it the mighty, though for the most part shallow and treacherous stream it is. For some hundred miles, and higher up than where I was, the banks of this river have no wood, but they are clothed with the prickly pear, and there being in those localities no growth of timber, the waters are free from the terrible snags which render the rest of the navigation so dangerous and difficult. The last "city" on the Missouri—a very few tenements or huts are often dignified with that at all events civilized appellation—is the "Sioux city." After that city a line of forts, held by the troops of the United States, is supposed to keep the red men in check, and assure the interests of trade. That these forts, though to some extent they open a line of communication, do not keep the red men in sufficient check, I have had an opportunity of knowing, as will appear in subsequent narration.

The first meal which I partook of on board the "Sky-

lark " was a liberal tea and supper, and of course as there
was a very large number of people on board, I prepared
myself to observe if the belief of my friends in England
was correct, as to the rush of rude and hungry people to
the chairs at the table, and as to their seizure and
manual demolition of any dishes that were near them.
Guess my astonishment, then, when I saw no rush, no
undue haste, but the whole company take their places in
the most orderly manner, and what was excessively to
my delight and comfort, no man spit while sitting at
the table—and wherefore ? Why, simply because the
American nation must take the quid of the United
States out of their mouth while they adopt the action of
mastication ; the quid being usually so immense, a
man could not eat with it in his mouth. Not only was
there no rudeness and disorder, but everybody who sat
near me, of whatever class, for we were all mixed up
together, was as courteously civil as could be desired,
and I began to suspect that my countrymen in England
knew as little of the real state and station of the Americans on board their river steamers, as Dickens, Mrs
Trollope, or Mrs Beecher Stowe did of the real state of
the slave ; or Messrs Cobden and Bright of the value of
railway speculations in the United States, or of the intrinsic worth of universal suffrage and the ballot.

When night came on, of course I expected to see
lights out at given points of the river to guide us on our
murky way, but there were no buoys, no lights, no
landmarks that I could detect ; yet the steamer moved
safely on as if by instinct, and I began to marvel at
everything I saw, or did not see, or, in fact, at my
immediate position. On going to bed, then, indeed, I
found discomfort, for, in addition to the warmth of the

weather, the hot steam or vapour, for it was not damp, found its way from the engine-room into my cabin, and produced such a state of thick and heated atmosphere that respiration was impeded, and sleep out of the question. In addition to this, the thumping of the machinery and roar of the wheels were close under me, and altogether they produced such an amalgamation of noises that I wished myself in my waggon on the upper deck. To remedy the evil, however, I shifted from the lower to the upper berth in my cabin, and set open each little window above the respective doors, and thus from time to time caught a mouthful of fresh air.

Morning came, when, having bribed a sable, and therefore civil, attendant to fill my blessed tin bath trunk with the cool waters of the river, a giant refreshed, I stepped out on the lower deck to worship the skies, and to revel in the fresh but not the sweet air that came through the autumn-tinted woods. When I say "the fresh but not the sweet air," I speak with comparative reference to the airs of Old England. In America there is not that sweet diffusion of fragrant flowers or withered bloom in the atmosphere which so often reaches the senses in England, and speaks to the soul of persons and of places with whom or in which bygone hours of happiness have been passed, for the wild flowers of America have not any of them a perfume, and such scents as the withered leaves might be supposed capable of affording are disguised by a palpable or earthy smell of dust, and they, like the flowers, tell no tales of a summer hour. The morning was lovely, and the sky above the forests—that stood on the banks of, and were tumbling into, the river, as if sending other trees to swell the numbers of the pallid and drowned but still dangerous giants of a bygone generation already fallen in—was pink and grey at first, and then beautifully blue. The heavens as well as the woods looked down on the rippling stream in strong contrast to the bare, brown, and sandy shoals, and gaunt limbs of trees which rose above the waters, bleached and fixed, as if their outstretched arms, once raised for rescue, had been immovably petrified in the action of a dying prayer.

The safe manner in which we had been steered in the hours of darkness, and the innumerable snags and shoals above and beneath the water that intersected the course of the river, made me look up from the higher deck to which I had ascended in admiration to the still more lofty pilot-box or place of steerage whence the captain directed the way. He was not there, however, being for a time off duty, but I saw him seated on the deck close to me. The salutation of the morning having been made, I took the opportunity of expressing to Capt. Sousley my admiration of the river steerage, as well as my thanks to him for all the comforts I found, and that had been provided for myself and my dogs on board his ship. His conversation showed me that he was an intelligent gentleman, and one well versed in his difficult duty, and resolved with all his energy to attend to it. In addition to this, he was a sportsman and a keen observer of the habits of wild animals; and for the remainder of the voyage at times, by night as well as day, I was with him in the pilot-box, which, being surrounded on all sides with glass, was not only the best place for seeing everything, but it was there that I could form some estimation of the extraordinary knowledge of the river possessed by the captain, and the way in which the one sole man governed progression and cared for the safety of us all. The paddles of the "Skylark" worked independently of each other, and there were bells in the pilot-box within reach of the captain's hand as he stood at the wheel, which ruled the pace, suggested more or less steam, backed her, or put but one wheel in motion at a time.

As we were steaming up the river, the snags, imbedded in a shifting and sandy bottom, all lay *chevaux de frise* fashion, with their sharp arms invariably tending down stream, the bolls of the huge and dangerous impediments out of sight, but the "snags," or points of destruction, peeping above the water in several places wide of each other—many other limbs, of course, ready to catch a luckless boat, which, from being beneath the surface, were not visible to the eye. Not only, then, were there these visible snags, but of course there were hundreds of others, all equally fatal to the side or bottom of a vessel, that lay concealed; and what was worse, the captain assured me that they kept shifting their positions, and he never knew for two voyages together where to place them. Sticks, scum, floating wood, and the *débris* of ruin and death often caught on these pointed snags, and sometimes their pallid hands arrested ghastly things as to the fate of which there was neither policeman nor coroner to inquire. Thick and muddy, or, more properly speaking, sandy, as the waters of the Missouri are, they are, nevertheless, deemed to possess the most healthful qualities. The Americans say "they are better than a doctor"—no great meed of praise, if I am to judge of the young men who minister to disease, or rather unto death, in the frontier towns. For myself, while on board the "Skylark," I never felt in better health. It was not long before I saw (on Saturday, the 19th of September) a pretty fair sample of the fate (but for the greatest skill and caution) that might have happened to us; for, on my exclaiming, " A-ha! what is that interesting object ahead of us ?" (we were then just off Augusta Bend, a little town on the banks of the river), the reply was, " Oh, that is the wreck of the ' Duncan Carter,' the steamer that stuck on a snag about three weeks ago ! guess the snag went slap through her bottom, and held, and still holds her, like a vice."

On the following day (Sunday) I had a specimen of a thunder-storm in these latitudes ; it came on in the morning, preceded by a high wind, driving a fog-like cloud of dust before it, and put me quite out of conceit with English thunder and lightning, giving me some insight as to what I might expect to hear and see in that particular when on the prairies of the Far West. It was on this day that my eyes were intensely gratified by the sight of the first flock of wild turkeys I had ever seen, as they sat after the rain to sun themselves and dry their bronzed feathers on a sand-bank beneath the woods. After the thunder-storm had ceased it became very hot, and at night when I sought my berth I found it so close, full of steam, and uncomfortable, that to attempt to sleep in it was impossible; so I resolved to try a night on the cushions of the seats in my ambulance waggon. For that purpose I repaired to it, carrying with me a blanket or two and a pillow, when, drawing the outer covering to the tilt as close as possible, I went to sleep. At about midnight, not only an acute sensation of chill aroused me, but also a rushing noise of the wind in the rigging of the vessel, and the unpleasant trembling of the waggon itself put me very wide awake. It was, in fact, suddenly blowing the first of the equinox, and the temperature of the air became all at once very cold; in addition to this, there was a very vexatious sensation as to the security of

my selected sleeping-place, for the confining rope was
single and rather slack, and it seemed in no way improbable that I might go from the upper deck, waggon and
all, into the waters of the Missouri. Having descended
from my waggon, I found the mate on the look-out ahead,
so from him I obtained an order for the better security of
my waggon; this done, having sought my berth, then
indeed I rejoiced in the warmth it afforded. To show
the state that that berth was likely to be in in hot weather,
I have only to say that the floor of it was as warm as a
toast, from the vicinity of the engine-room beneath—no
uncomfortable thing that on the night in question.

The foliage, as we ascended the river, every hour increased in the autumnal tints and still more varied crimson, scarlet, and gold, as well as in the ferruginous reds
which now so frequently intersected the masses of green;
rocks also occasionally margined the stream, and promised
to afford dens or earths to bears, wolves, and foxes.
The whole way along the river I had opportunities of
observing the graceful flight of hundreds of the turkey
buzzards. On one sandbank they had found something
which had been alive; but it was so devoured that I could
not be certain as to what it had been; from it they idly
soared away, winged aldermen from a feast of carrion,
and then, perched on the bare boughs of old dead trees,
with outspread but stationary pinions, they invited the
sun to assist to dry their feathers of the filth and promote
the interests of digestion. From the sides of the river
there also frequently arose small flocks of the common or
English teal, and everywhere the common summer snipe
and sand-piper abounded. During the voyage we passed
some most fertile-looking land, still known by the Indian
name as the "Wakendah" Prairies; the effect of the

wild sunflowers which studded the grass of these plains was bright and beautiful in the extreme. For the first time in my life I also saw a flock of pelicans on the wing, so that I began to think I was at last approaching the longed-for haunts of the wild game, and that it could not be very long now before I should shake myself free of the habitations of men, emancipate my dogs from every durance vile, and with them seek and share the blessings of the boundless desert.

The society on board the "Skylark" was not only very extensive, but mixed and perpetually changing—in short, the steamer inclosed a little world within herself, and gave full opportunity for the study of American character in all its phases—from the well-mannered gentleman down to the Boh-hoy, card-sharper, and false lottery keeper. From the latter personages I derived on this occasion intense amusement. From the first moment of my setting my foot on board the steamer, and throughout the voyage, numbers of men sought my acquaintance and proffered their advice, and gave me some trouble, but not much, to distinguish between the real gold (there was a good deal of that) and the dross that attempted imitation. I have great facility in reading character, and somehow or other there is always a spirit within me that inclines me to the good man, however poor his exterior, and gives me timely warning against the bad. Thus I soon began to read my river acquaintances as if they were books, and in doing so was immensely amused. I had not been long on board before I discovered that cards among a great many were a favourite occupation, and that noisy groups engaged in them, who I thought laughed a great deal more than was natural or necessary, and therefore of course the idea suggested itself that they did so simply to attract attention. One thing in regard to a highly-bearded individual also struck me as peculiar. On whichever side the ship I went, as well as perpetually near the door of my berth, which gave admittance to the lower deck, there this short, stout, hirsute fellow was, with a portable table in front of him, and a pack of cards in his hand, with which he always seemed to be playing alone and against himself.

If I changed my position on the vessel and took a chair on the port or larboard side, so sure was I to see a table and then a stout man accidentally pushing themselves round the bluff entrance to the saloon, till by degrees both were well before me. Of course it soon became very evident that this card-sharper desired to engage me in a game at cards, under the, to me, flattering supposition that an "English nobleman rolling in money" would play at any game with a man of whom he knew nothing. On observing that this broth of a Boh-hoy was such a fool, I thought that I would amuse myself with his folly, instead of permitting him to enrich himself with mine, and I let him think he had caught my eye or fascinated me to destruction by keeping my glance occasionally or furtively on the cards while he was at play. I saw that he observed this, and for that reason haunted my whereabout with his table more than ever. The encouragement I gave him, however, was very slight, and it was evident he did not feel quite sure of his game, for not until the third day of the voyage did he venture to address me. On that day I had a chair on the lower deck, just outside my door, when I saw the everlasting table coming round the bluff entrance to the saloon, followed by the stout man, lugging after him a chair. He

did not like to intrude his rascality too near, I suppose
from an uncertainty as to how I might take it; but having got within ear-shot he commenced a game with himself, and after much frowning, and many a deal, and im-
patient " pshaws," he exclaimed, as if to himself, violently
rubbing his thick red hair, " Pshaw! it's no use for me
to try, I *can never win*." On this exclamation I fixed my
eyes on his cards, when, thinking the golden opportunity
had arrived, he said, in rather a dubious tone, for I believe he suspected that I was wide awake, " Won't you
try a hand?" " No, sirre-e-e-e-e," I replied, when the
fellow, table, and all vanished round the corner of the
saloon, like a harlequin in a pantomime. For the better
information of my English readers, I must explain the
force summed up in the concise method of my refusal.
In American parlance, to say " No, *sir*," laying a long
emphasis on the last syllable, is to give a contradiction
of a *very flat sort;* but to say " No, sir-ree-e-e-e-e " is to
convey the sentence of " No, you infernal rogue, I'll see
you at the devil first; " and therefore my method of
speech being fully known to the card-sharper, thus he
vanished. I only saw him once after, and then he lay
listlessly in the hands of the ship's barber—a very good
one—for the purpose of being shaved and clipped.

From a subsequent conversation with my servant I
found, also, that some of the crew of the vessel had
warned him that there were four or five card-sharpers on
board, and to him some of the gang had addressed their
attentions, but in vain. George's watch over the safety
of my dogs was very well arranged, and it was settled
that whenever the steamer stopped, either to put ashore
passengers or to take in freight or wood, he should be
with his dogs, and I should be on the look-out on deck;

but my opinion is, that had the dogs been left to themselves, their entire safety would have been insured by the
vigilance of Brutus and Druid. The day before we
reached Kansas a free black made very free with the
head of as sable a shipmate, and nearly cut it off. It
was in this voyage that I made the agreeable acquaintance
of Mr W. L. Harper, to whom I promised to send out
from England a couple of fox-hounds.

The panorama of the voyage, in some places, was interesting and pretty, for whenever we approached a town
or village the band on board the ship struck up a lively
air, and then the people were seen running from their
houses to meet us. At one lonely hovel, where we took
in fuel—wood being in such plenty, and costing nothing,
it did the duty of coal—we (myself and many of the passengers) went on shore and gathered the " pau-pau "
fruit from the shrubs on which it grows. I found it delicious, but my companions assured me that it was thought
very little of in the United States, and they told me, with
some *naïveté*, that " pigs would not eat it." I ate it,
however, and enjoyed it very much—so much, that I
brought some of the seed home with me, and, with some
other seeds, sent them to the gardener at Taymouth Castle to see what they will do in that soil. On the banks of
the Missouri there are several settlements exclusively
German, and in all these is grown the grape that makes
the "Catawba" wine—the wine I allude to is so pronounced, but I am not sure if I spell it correctly. If made
from a good vintage it is excellent, and to my mind,
when good, of a finer flavour than champagne. The
best I tasted was at the Planter's House at St Louis; it
was slightly up when uncorked, and when properly iced
that sample of the wine was delicious, and my opinion is

that it would take very well in England. Captain Sousley informed me that beavers and otters were to be found
in all the tributaries to the Missouri, and I subsequently
heard that since the demand for their skins had decreased
those sagacious animals, the beavers, were becoming infinitely more numerous than they had been of late years.
Captain Sousley also told me that on a still summer's
evening, by sinking the wind, he crept near enough to a
beaver dam to see the animals at work in cutting down
trees, and that it was perfectly astonishing to observe with
what rapidity they got through the substance of a good-
sized tree, and then, when it tottered to its fall, how they
congregated, just as men would do, on the safe side, and
reared themselves on their hinder legs to watch the effect
of their labours.

On the last day of the voyage, one of the *gentlemen*
who had been very civil and attentive to me, instructing
me in the ways of the New World, and on all subjects
save that of a lottery, came out in a new light, displayed
a collection of tin and tinsel, and some very suspiciously
wilful dice—which had a great inclination to rest on particular sides, and with which the lucky caster was to win
" tremendous bargains." Of course I threw for nothing,
and while looking on I had to ask undoubtedly good
American gentlemen who did " put in" how they could be
such geese as to be so deceived ? They replied, laughing,
that they knew all about it ; but they were willing to lose
a little money for the purpose of passing the time. I said
then, I would have seen who could throw a dollar farthest
into the river, rather than let a lot of thieves think they
had "done" a gentleman. How I laughed in my sleeve !
but nevertheless under a great inclination to stride up and
kick over the table, when I saw a fellow, whom I had

observed to be an accomplice of the card-sharpers, go up and win the best lot on the lottery table, and then, when the presiding fool and rogue bid him a considerable sum for the prize to be returned, answer his bosom friend rudely as to "not taking twice the sum named for the treasure the dice had given him." "Oh," I thought to myself, "these American Boh-hoys *will* be the death of me!" but I refrained from any just interference, and let the farce go on.

But hark! the band strikes up a lively air; there is an unusual bustle aboard; people crowd the upper deck, and we are close in on Kansas city. Men and boys run from the houses to the quay; we shorten speed, and the steamer brings her bluff bows alongside the land; we collect our luggage, and strangers board us to welcome friends, and learn the latest news.

I had gathered together my trunks and boxes, and had piled them one above the other, a favourite little black shining valise of mine, containing my hunting materials and any little loved remembrance of home that I had with me, on the deck at my feet, and my dogs still safe in their kennel, when several Americans, and one young Englishman, came to me and expressed their compliments and kind congratulations on my arrival. The first suggestions that reached me were as to the two hotels, then offers of service and advice became so general, that I stood in doubtful silence, simply on guard over my property, resolved to take no step of any sort till the conflicting speakers had tired themselves and left me more alone. While thus stationary, the crowd seemed to give way to an important personage, and a finish-looking, half-cast, Indian-eyed man came to the front with rather a theatrical air, and somewhat to the following effect delivered himself: " I,

sir, am the chief of many war tribes. I saved the lives
of Lord Stephens, Lord Brown, and Lord Thomas John-
son, and other English noblemen, when the red warriors
had doomed them to death. I am, yes, sir, known as the
Bloody Arm of the Rocky Mountains; yes, sir, and my
deeds have gone forth, trumpet-tongued, throughout the
universe; and I shall be happy, yes, sir, to protect you."
This was said with so boastful an air, that I very nearly
burst out laughing in the dark man's face. While this
dark warrior strutted his hour, a very mild but excited
little man, close under my right shoulder, "took up the
running." "Yes, sir," he commenced ; "Lord Berkeley,
I *am* happy to welcome *so* distinguished a visitor *to* our
wonderful land—the almighty gem, sir, of the western
hemisphere. Yes, Lord Berkeley, and if there *is* anything
I *can* do to promote your wishes, name it, Lord Berkeley.
Yes, sir." "Squit," and, with a curious aim, he spit
right into the beloved keyhole of my little perfumed valise;
it had been given to me by a lady, and therefore kept
unsullied, and on this valise, in a nervous way, he fixed
his furtive glance. "Thanks, sir," I replied. "Squit"
went my friend again slap into the same keyhole, and I
did not know what to do. "Thanks, sir," I repeated,
" I will tell you if I require your services." "Squit" went
this man of unerring aim again, and while he was thinking over a further address, "squit" he went again into
the same assaulted keyhole, which I suppose by its
brightness had fascinated him, when I resolved to interfere. . Not wishing to hurt his feelings I paused for an
instant, when " squit" he went again into the very same
spot. "I beg your pardon, sir," I said, " perhaps my
favourite little valise is *in your way*," reaching my hand
out to remove it. "Squit" went the persevering lip-

rifleman again, narrowly missing my hand; "Not at all, sir ('squit'), it's not incommodious at all," and "squit" he went again, delivering a double shot this time. This was more than I could bear, so pushing away my valise with my foot, I said, "Sir, my writing materials are in *that* little trunk, and if your tobacco juice gets in anywhere my paper will be considerably soiled." "Good heavens, sir," said my little nervous friend, now covered with confusion, "I beg ten thousand pardons; I did not know what I was doing; I am quite shocked, Lord Berkeley," when looking really the picture of remorseful and mortified distress, he searched his pockets for a handkerchief, *but he had not got one,* which, as he saw that I was observing him, increased his confusion. Being sorry indeed to see my little friend thus annoyed, for he had intended me no discourtesy, I espied on the deck a piece of an old "swab," which with a glove on I instantly seized, and with it rubbed the filthy stains away, assuring my friend that it was all right, and no harm whatever done in any one way.

By this time "the Bloody Arm of the Rocky Mountains," and many other people and idlers who had come on board, had exhausted themselves and taken their departure, and I then accepted the invitation of the gentleman who kept the Planter's House in Kansas city, to take up my residence with him during the time that I was perfecting my camp equipage for the desert, and I did so chiefly because his house of entertainment was a little out of the town, and, in addition, he offered me good rooms for my servant and my dogs. The Englishman, Mr Powell, who had accosted me, also having offered me his assistance in any mode I might require, I at once accepted his proposition with all thanks, and it was with great satisfaction that I then learned that his father, living at Hanham, near Bristol, had been a constituent of mine while I represented the Western Division of Gloucestershire in the House of Commons.

On arriving at the Planter's House I found it a low building of one story, and that my room was therefore on the ground-floor, while, in some storehouses near, George and my dogs obtained very good accommodation. Thus, then, I had at last arrived at Kansas city, the extreme point of civilization on this route to the prairies, and with the greatest alacrity I became, not as "Japhet in search of a father," but as an English hunter in search of and most anxious to obtain not only seven good mules, but three able and active horses, fast enough to carry fourteen stone up to a bison going at the top of his speed, and a pony from whose back, while on the march, I might obtain all the smaller game that chance, or "Chance," indeed, might enable me to discover.

CHAPTER X.

I HAD not been long in Kansas city before I began to
find out that I had indeed left all the delicacies or luxuries of life behind me. Ice, for which I had hitherto
found America so famous, and in which great comfort I
had revelled from the moment I had set my foot on
board the " Africa," was no longer to be procured; the
last fraction of it had gone on in the "Skylark" to the
further limit of her voyage, and with the ice the art of
plain cooking and tolerable viands. Brandy, whiskey,
or sherry, if asked for, were to be procured *by name* only
at the stores, and, when procured, the former two were
coloured spirits of wine, and the other a strange concoc-
tion of inventions, not one of the component parts of

which in any respect whatever had anything to do with a vine. Coffee was bad, and tea worse; and, as to cooking and quality of meat, it was as if the Fiend himself had toasted a lean and incorrigible sinner, and sent him up not in his general humour, but for once to scare his fellows from coming to his fire. If a fowl was killed, it died without reference to age, and was sent to table the same day. Pastry, water-melons, and pumpkins, some very good sardines and eggs, were there the only wholesome things to be procured; and some of this fare so much demanded strengthening, that, with a heavy sigh, I had to break in on a flask of whiskey of the better sort given to me by a friend on the railway. It would not have been wise, of course, to intrude on my stock of brandy and wine, or I might become short of it when most needed in the desert. Mine host and his pretty daughter did all they could to make me comfortable and to amuse me, and I was very well waited on by a man of colour as well as by a white waiter, but the crickets and beetles that came to inspect me in the night, and to hold *levées* on the floor of my room, *nomen illis legio*, and I could very well have dispensed with their attentions.

My first step at further preparation for the prairies was to order a covered little four-wheeled waggon, to be drawn by one stout mule, for my dogs; the ingress to it to be from behind, and the door on hinges, at the bottom of the cart, barred across, to make them a sort of stairs to get in by, when the door was inclined by being supported on a staff. Beneath the driver's seat in front was a little box to hold anything that might be required. The waggon I had made out of old materials—very old; for the maker of it, with true coachmaking craft, for a customer that he might never see again, did not fail to put in all the otherwise useless stuff he could thus get rid of. Precisely like his fellow-tradesmen in America, this waggon-maker set about his task with perfect indifference as to when or how he should finish the article required, and I found that he paid very little attention to my repeated exhortations as to haste. However, by dint of repeated visits he put the body of the waggon together, and of that fact I made a spur, for I assured him that if he did not finish the waggon by a certain day, I would go away without it, and he would have worked, so far as he had gone, for nothing. The next thing I did was to ask Mr John Campbell, who resided there, to whom Mr Robert Campbell had given me a letter, and Mr Bartley, a general merchant, as also Mr Powell, to recommend me a guide for the plains—a man, in short, to whom I could safely intrust the contents of my camp, and who was so respectable that he might be permitted to hire me six good men, who were to act under his directions. I told these gentlemen that if they could find such a guide and trustworthy man, they might also fix the wages I was to give him, for, of course, if such a man could be found, he would be worthy of his hire. They fixed on an individual named John Canterall, one of whose hands was deformed, and how he turned out, or what that rascal's real value was, remains to be told, the inordinate wages for him having been fixed at one hundred dollars a month.

The first thing this fellow did was to assure me that I had not sufficient stores even for himself and six men, and that in his great conscientious duty he thought it right to inform me that, with the Indians in a disturbed state, and the number of half-starved and lawless emigrants returning disappointed from Pike's Peak and the vicinity of the gold fields, he felt he could not be answerable for the safety of my camp, horses, and mules without two additional men. To this plausible tirade I replied that, as two more men were all he required, and as I wished to have all trouble taken off my hands, in order that I might give my mind up to sport and observation, he might have them, and thus it was settled. The dog waggon, bell tent, &c., capable of comfortably containing nine men, with some additional stores and extra fittings within the heavy baggage waggon, were then ordered, and I gave Mr Canterall—Cantwell would have been his more apt name—instructions to explain as follows the conditions on which I would hire the men:—Their wages were to be twenty-five dollars a month each; any man displeasing me to be paid off at once, and anywhere. They were to bring their own arms and their own blankets, and to take care of and watch my camp, and distinctly to understand that, as I intended to do all the sporting part of the journey myself, they were simply hired for work. I heard this fellow Canterall, *while I was within earshot*, explaining these orders to one of the men very well; but shortly after that he came to feel his way with what he said was "the request of his men," and he asked me "if they might all bring their ponies?" To this I answered very sharply, "*No*, my horses and mules would be quite enough for them to watch and take care of, and therefore I would have no animal in the camp that was not my property." He then set about packing and arranging the baggage waggon, while I instituted inquiries for horses and mules, and expressed myself desirous of having such as were for sale brought for my inspection.

Of all the wretched animals I ever saw on the same

day, the quadrupeds, horse and mule, that were at first brought for me to look at were the most worthless, while, at the same time, their owners demanded prices at which I absolutely laughed. "Do these people take me for a fool?" I asked of one of my friends, "that they think thus to cheat me?" "No," he replied, "not exactly that; but they judge of you by a countryman of yours who preceded you here, and who bought anything that was recommended to him, and, after buying, changed his mind and sold, and then bought again; so they hope to make something in a similar way out of you." "They will find themselves considerably mistaken," I said, "and, to make short work of it, you had better go about among these people, and tell them that so disgusted am I with the sort of animal shown me in these parts, that I have nearly resolved to 'still hunt,' or stalk the buffalo, and take no horses at all." On the following day I gave strength to this rumour I had thus desired to be promulgated by refusing to look at any of the horses that were brought to the Planter's House.

During the cessation from business that the execution of my orders, and my pretended resolution of not buying any horses, afforded, I contrived to amuse myself by looking at the wilderness, which, with its wooded scenery, came close upon the verge of Kansas city, and by an excursion or two to villages in the vicinity, which latter were, in fact, the last haunts of collected men, ere I broke out into the wilds. The road from the town to these villages through the woods was a very primitive one, from which the trees had not been entirely taken, but left standing to some height in the roots, cut off just high enough for a pair of horses or mules to "straddle," or, in other words, to quarter or go each side of, permitting the

stump of the tree to be between them, and to be passed over
by the body of the carriage and between the wheels. In
a dark night this road must have been impassable. Among
the many kind civilities shown me in America, a very pretty
girl of about 14 or 15 was once permitted to drive me out
in what her father called his buggy, but which carriage
had four wheels, and was a sort of curricle drawn by a
pair of horses, to show me the woods in the full effulgence
of their autumnal tints. The afternoon to which I allude
was very hot, and the declining sun giving signs of a
glorious set in purple and in gold, and the air so hushed
that on the road (or no road) in which we were, there
was no noise by horse-foot or wheels, and therefore no
difficulty in conversation. The pretty, smiling girl was
driving, with a mare on the near side, very free to her
work, and hot, while on the off side there was a horse as
lazy as anything I ever saw. My charming companion
was perhaps a little diffident of being able to amuse "an
English nobleman," and consequently a little shy and
nervous, but still most anxious to keep up the conversation.
She kept looking up in my face with her bright, laughing,
yet half-timid eyes, while at the same time, with a sort
of feverish twitching of the whip in her hand, she never left
off touching up the mare, who had already done the entire
work, and driving her half mad. The mare at last, lather-
ed from head to foot, while the horse was as cool as a cucumber, began to give in, so I said, "I will tell you what,
dear, if you keep aggravating the mare in that way, you'll
knock her up."

"Possible! well! hum—hum, I'd like to know!" replied my companion, more bashful than ever; "but, don't,
Lord Berkeley; you—you oughtn't; don't, Lord Berkeley."

"Don't!" I exclaimed; "yes, you will, if you keep

fidgeting her in that way with the whip you'll assuredly knock her up."

" Possible ! " rejoined my dear little friend; " but you shouldn't—I'm sure if I thought so I would never touch a whip again. Don't, Lord Berkeley—you oughtn't— don't—"

Looking into her pretty face, wondering what on earth it was that I could have said, I saw that she was blushing and really embarrassed; so I changed the conversation, amused, though, by observing that my caution as to the mare was attended to, as my friend carried her whip more erect, waving it after the manner of a fairy's wand. Our most agreeable drive terminated soon after this, and, as I handed her from the buggy, I saw she had grown so laughingly bashful that she dared not permit her eyes to come in contact with mine, and I continued to marvel at what it could have been that I had said to make so inexplicable an impression.

In the evening her mother had a tea party, at which she presided with a long clay pipe in her mouth, an indulgence frequently luxuriated in by the married ladies of the frontier towns and villages, when after tea she put out her pipe and got up to stand by the door " to receive company." Of course I took my place near her, with a view to observe American frontier usages and manners. Two or three guests arrived, and then a tall, high-shoulder-ed, raw-boned gentleman, with a long nose, small eyes, and a high forehead, very upright, very stiff, and very grave, his face decked with a long pointed beard, stained either by the sun or the juices and smoke of tobacco, and redolent of spitting and of pipes.

" Good evening, Mr Smith," said the hostess; " but where is your good lady ? I had hoped to have seen her with you."

" She's knocked up," replied the matter-of-fact guest.

" What! dear me, *again in that interesting situation ?* "

" Oh, good gracious! " I cried to myself, as I turned away to conceal my great inclination to merriment; " now I know what it was that I said to my pretty companion in the buggy. 'Well, did you ever ?—no I never;' hang me, but I will get an American dictionary, and if there is not one in print I'll publish one."

The good effect of my pretending that I would not buy horses, on account of the miserable and extortionately priced quadrupeds that had been shown me, was soon made manifest, for there were several animals of a better shape, and at less money, brought to the door of the Planter's House.

" Well," I said to one of the riders, " let me see the action of your horse; " when, holding fast on to a single sharp bit, too tight at the checks for comfort, the animal's head imprisoned by a needless martingale, and half strangled by a throat-lash strained up to the highest hole, I saw pass before me a horse with the most disjointed, agonized, and inexplicably complicated action I ever saw —so unnatural and complicated that I could not discover where the ailments in nature were.

" Well," I said, " that is the rummest brute to go I ever saw "—I used that homely phraseology because it suited the men I spoke to — " he's lame all over." " Lame! " cried the rider, " guess you're pretty considerable out *there;* thought you'd like a ' pacer;' hoss can go anyhow, pacing, loping, trotting, shambling, niggling, ambling, or what not, and run slick over all the buffaloes on the ' peraries.' " (It is the custom of the country to pronounce the word prairie as if it was spelt with three syllables.) " Well, then," I replied, " let me see the

horse 'lope' as you call it, but gallop as I call it, and
trot and walk—in short, put him through his natural
paces, and leave these new-fangled agonies of a more recent world alone, they are of no sort of use in the chase."
Thus apostrophised, my dealing friend did as he was desired, but with very great difficulty the horse was kept
in such a constrained position. "Here," I cried, to the
sable attendants of the stable of the hotel, "take that
horrible unicorn saddle off that horse, and put my English saddle on; take that gagging bridle out of his mouth,
put mine in, and toss away the martingal, and I'll throw
my leg across the horse and see what he will do with me."
The emphatic way in which I spoke of these things greatly
amused a knot of my friends who were looking on, and
while the horse was gone into the stable one of them said,
" Guess you won't try riding on that English saddle
o'yourn arter buffaloes ? " " Guess not ! " I replied.
" Why ? " " 'Cause if you do," continued my friend,
" you'll come to considerable difficulty. What have you
got to hold on by gin the beasts make you turn, and
mighty sharp ? " " My knees," I said ; " and when one
of my hands is filled by my carbine, and the other by
the reins, what the deuce do I want with that stupid may-
pole of yours, unless to hang my hat on ? " This rejoinder again intensely amused the by-standers, and the
horse having been brought out equipped as I desired, I
put him through his very good paces, and showed them
that with my double rein the horse never put up his head
too much, nor pulled an ounce, though I had to kick him
out of his false action whenever he put it on.

After many horse-dealing scenes of this sort, I selected
a chestnut horse, but which they called a sorrel, with

11

whose action and mouth I was delighted, also a brown and
very well-bred looking mare, whose owner was a very nice
and fine young fellow, and who knew more about the
real merits of riding than any other man I met in Kansas
city. He asked me to pay him a visit on my return,
which offer I regretted I could not avail myself of, as I
returned by another route. The beautifully smooth action, mouth, and temper of the brown mare Sylph (the
name I gave her) endeared her to me much, and, could I
have contrived it, I would have taken her and the chestnut horse, whom I christened Taymouth, in recollection
of my visit to Lord Breadalbane, to England with me.
Sylph had a very bad blemish as far as appearances went,
from having been blistered too severely on the withers,
where a fistula had evidently been apprehended. When
I was looking at her with a view to buy, the man who
showed her to me accounted for the eyesore on the wither
by saying " she had got that blemish by jumping slick
through a window." Rather a 'cute assertion this, but
difficult of belief—in the first place, because the blemish
was perfectly even, and as much on one side of the withers
as on the other; and certainly, in all the hovels called
stables in America that I ever saw, nothing larger than a
cat could have jumped, or would have attempted to jump,
through the vacuum which their apertures, called windows,
afforded. The third horse bought by me was a strong useful bay, in very good condition, not so well bred as the
other horses, not so fast, but with speed enough to get up
to a bison; and this animal I intended to carry the fellow called a guide, and to be used by him to trot on and
fix the ground on which we should from time to time encamp. I regarded this horse as my hack hunter; the
other two were exclusively for my own riding. To this

stud I added a little pony, as fine a goer as man ever saw,
and as quiet, good-tempered, and sensible, from whose
back I intended to shoot the smaller game on either side
the line of march. Poor little Charley, as I called him
—ho too had been taught all sorts of false and tricky actions, and it was a long time before I could disabuse him
of an abominable amble, and reduce him to the natural
and shooting pace of a steady walk. For the chestnut
horse Taymouth, sound wind and limb, and free from
blemish, and fully able to carry me to any hounds in
England, I gave the sum of 225 dollars, or £45 English
money ; for Sylph, the brown mare, the same sum ; for the
bay horse, which was in the best condition, 175 dollars, or
£35 English ; and for the pony, 40 dollars, or £8 English.

After the horses were in my stable, my American
friends suggested that I had given more than their value ;
but I was quite content, for I well knew that nothing
could be better than the action they all possessed ; but as
to their facing buffaloes, that I knew I could not expect
horses that had never seen them at once to do. In passing through the town after these purchases were made, I
stopped to see Mr Powell, at what is considered the best
hotel, and while treating an acquaintance to drink at the
bar, as is the custom of the country, I was intensely astonished and amused to hear that the gentlemen then
and there present *had made a match* with my horse and
mare, Taymouth and Sylph, to run them one against the
other for two miles, and that without the slightest reference as to whether I would consent to it or not ; betting
was considerably brisk. Let it be also borne in mind,
that I had pronounced them both in too low condition,
that they required nursing, and that I should start for the
plains at the earliest possible moment. With consider-

able glee as well as excitement my friends ˉproposed this match as a matter of course; but I said that, if it was all the same to them, it would be infinitely more agreeable to me, and better for my horses, *not* to give them any additional work, and therefore the match *must* be off. They received this staid announcement very well, and we parted excellent friends.

While at this hotel I had my hair cut, to make it fitter for the dusty plains, and when undergoing the operation from as able and sable an artificer as ever I saw, one of his assistants, also a youth of colour, took a guitar and sung a plaintive ballad, with as nice a voice and as good taste as ever I heard united. I was so pleased with him that I offered to take him as my minstrel and cook (rather opposite appointments), and he was very willing to come, but on consultation with his friends he was prevented. I take it, if he had absented himself, he would have broken up a trio of negro melodists. Nothing could be cleaner or more *recherché* than this barber's shop, contained as it was in a spacious apartment within the hotel, its walls decorated with pictures of slightly questionable propriety, as to which I seriously apostrophised the sable operator, much to his amusement and that of his two assistants. I insisted on my hair being cut much shorter, but in its usual fashion, while he struggled hard to clip it in a new one, when, as he was inclined to be unruly, I solemnly assured him that if he turned me out in any new fashion, nothing on earth should prevent my hitting him on the string of his apron, and leaving him doubled up until my return from the prairies. Under this threat he made just so much change as would, and did, take a month to put right again; but he was not doubled up, because I did not discover the aggression until the next morning.

My *ruse de guerre*, in regard to a better class of horses being brought me, answered very well; but a similar threat to use oxen instead of mules did not create a corresponding feeling in my favour, as my friends at Kansas city " guessed considerable " that I should not really proceed with a slow conveyance. While I was thus kept in suspense, intelligence came that there was a man encamped on the plains, a short distance from the city, possessed of three or four hundred mules, and that they were all on sale; so, ordering a conveyance, myself, Mr John Campbell, and Mr Powell, at once proceeded there to inspect the camp of the American muleteer, and I found him a very straightforward good fellow. We found his tent and waggons provided with sable attendants from Mexico, and ornamented with a very good-looking girl, I suppose the master's mistress or tent-keeper, as there was no house, and a number of long canvas troughs on an excellent principle, and supported by sticks, for the mules to feed from—a plan which prevented waste of corn, and gave to the black Boh-hoys an excellent sleeping-place or hammock. Having nothing of the sort when on the plains, I had to feed my horses and mules with the corn in ear on the ground, when much of it of course was lost, and in which position most crafty pigs, perfectly versed in camping interests, could in a few moments sally from the nearest cabin and, if not driven off, eat it all up. The immense herd of mules was at a little distance from their owner's camp, removed there to some fresher pasture; but they shortly approached us, preceded by one of their body with a bell on, to the sound of which they are collectively attached, and driven up by a Mexican who had been despatched for that purpose.

While we were awaiting their arrival, a tall, lathy, din-

gy, three-cornered figure approached, with a high-crowned
tattered hat on, and a scanty but uncultivated red beard,
bearing on his shoulder an immensely heavy, small-bored,
short-stocked rifle, of a fashion a hundred years since—at
least I should have judged it so in England. I saw at
once that my two companions eyed him with dislike,
and that, though they gave him a civil greeting, they
would much sooner have met him in broad daylight than
at dark, and have been able to have greeted him, than to
have endured a nocturnal approach from a fellow against
whom they were not prepared. The man came up to
me, and held out his hand as usual, which I took, and
then relieved him of his rifle, with a desire to look at it.
We entered at once into friendly converse, when, on seeing
a projection in his coat on either side his hips, I said,
"What else have you got there? As I am going to the plains,
I do not intend that any man should be better armed than
myself." The rover and robber, for such he evidently
was, turned a glance at Messrs Campbell and Powell,
when, observing that their dislike of his company had
removed them to a little distance, he whispered to me
"revolvers," when, on his at the same time lifting his coat,
I saw that he had one on either side his belt. "Well,"
I said, "you beat me to-day, because my fire-arms are at
home, but I would wager that even now I beat you in
knives." On this he drew a short scalping-knife from his
belt, while at the same time I produced from my shoot-
ing-jacket pocket a splendid deer-stalking knife, a foot
long, or nearly so, with a pointed and trenchant blade that
opened on a clasp, so sharp that I could have shaved him
with it, and he with glistening eyes confessed "that
therein I beat him."

I then returned to my companions, for the mules had by

that time arrived, and, with much kicking at each other, were all eating corn from their canvas troughs. The owner of the mules then told me to pick out those that I approved of. I wanted seven, and I proceeded to walk past and inspect them from the rear, attended by two sable fellows, each bearing a lasso, and here for the first time I witnessed the exactitude with which that coil of rope was thrown. One of these men had but one arm, and yet he used the lasso, in the limited way I saw it tried, to perfection. Having pointed out six fine large mules, they had the noose at once thrown from behind over their heads, and they were led out in a little group by themselves. " Those are the mules I like," I said to their owner, "so tell me their price ?" " I cannot afford to part with any of them," was the unexpected reply; " they are my best wheel mules for the heavy waggons, and I could not replace them. You will see they are wheelers by the marks on their quarters."

" Then, why," I replied, "did you give me all this unnecessary trouble ?—or, perhaps," I continued, "I can answer that question myself. You thought the Englishman no judge of mules, and you did not expect that he would be able to pick out the best ? " We both laughed ; when, at my suggestion, he accompanied me in an inspection, and showed me all the mules that he was willing to dispose of. From these I desired them to lasso six, which they did, and for five of them I paid 110 dollars each, and for another 135 dollars, and six better mules I never desire to see in harness. There was one, a young mare, that had not been in harness much, with action and figure good enough to have been driven in the parks in London, and, with her fine but hasty spirit, we were obliged to be very careful. Having completed my bargain, and seen one of

these Mexican men of colour lasso a pig on the plains for
my two-revolvered acquaintance, I ordered the purchase
to be immediately taken home, and drove back to the
Planter's House. Before I quitted the plains, however,
having heard that a "*judge*," I forget his name, had a mule
to sell, I went to *his cabin*, under a sort of impression that
I should see an august personage in a wig, or, at least, in
expectation of finding a lettered gentleman. Guess my
surprise, then, when I was introduced to a man of the class
of an English day-labourer, living in a cabin, as far as I
could see, with only one man to assist him, both "the
judge" and his companion suffering from the aguish fever
and the effects of tobacco, affording a fair sample of the
rest of the settlers with whom I was so soon to be more
widely acquainted. His mule was not at all the sort of
animal I required.

On returning to mine inn, then, I found that Mr Canterall had *at last* sent me up one of the animals which no
doubt he and his compatriots had been keeping back, in
the shape of the finest mule of the whole lot. This
animal I at once secured to draw my dog-cart, at the
same sum which I had given for the first six; and then,
my numbers being completed, to my infinite amusement,
mules and horses poured in, their owners but to receive
the civil assurance, instead of money, that they were all
a day too late for the fair.

While at Kansas city I struck up an immense friendship with my blacksmith, who was from the very first as
civil, well-mannered, manly a man as any of his class in
England; indeed, I know not when I ever met one in
his calling who suited my fancy more. It seemed that
this good feeling was mutual, for I could trust him—the
first working American I ever saw whom I could trust out

of my sight—to take my horses to his shop and shoe them without needless delay. His charge was moderate, and his work was good. He stood about a head and shoulders taller than I did, and was powerfully made in proportion, and he introduced me to his brother, still bigger than himself. They both were going to the plains, and were very much annoyed that they could not make their journey tally with mine, that we might have hunted together. Nothing would have pleased me more than this, as one volunteer is worth a hundred of those men that I took with me. Mr Canterall, I could see, had made up his mind to stay as long as he could keep me in Kansas city; but by dint of immense exertion I induced the tradesmen to work, and thus had all things finished and collected by Saturday morning, so that I could complete my arrangements and have my waggons up to my hotel that day, so as to make a start for the desert on Sunday. The weather was so fine, and time at my disposal so short, that to me one day was of the utmost consequence.

I was very busy packing my ambulance at the stores where my dogs were, which was close to the side of the public road, when a huge jingling thing, called in America a stage-coach, with four horses, and many passengers, pulled up, and a big Boh-hoy, seated on the box by the coachman, thus in a very insolently familiar tone addressed me : " Hallo, guv'nor, so you're a going to the plains, and I guess I'll go along with you."

I had already had much to find fault with that morning, and being very busy packing, my frame of mind was not such as to put up with insolence, so I replied, " Guess there's two words to that bargain." " How so, guv'nor ?" replied the fellow. " Why," I said, " guess

you won't go without my permission, and that you
won't have." "Reckon I shall," asserted the fellow,
"so guess I'll go on into town and get my fixings."
"You can do as you like about that," I replied, "but as
I don't like your method of speech, and as I intend to
be master and have none but men with me, I guess
you're not the sort of fellow I want." "Guess somehow I'll get my fixings," rejoined the Boh-hoy; "go
ahead, coachman," and off he drove.

On the following morning, before we started, this fellow returned and made his application again, saying
"he had got his fixings!" but I told him, as I had told
him before, that he should not go, and I left him behind.

Eight men having been procured, besides Mr Canterall, or the fellow called a guide, we left Kansas city on
Sunday, at ten o'clock of the morning of the 26th of
September, and then for the first few miles I learnt what
it was to travel on bad roads, and what places it was
possible for American waggons and mules to get through.
We stopped at a little village to get a few more things
which my useless guide had forgotten; and shortly after
the plains began to assume their native naked grandeur,
and to afford me some insight into the face of the land I
should soon have to gallop over, my first exclamation to
myself being, "Oh! what a country for an English
pack of foxhounds and a thorough-bred horse!" A
gigantic Leicestershire lay before me, without a thing to
stop horse or hound but pace or the death of the hunted
animal!

CHAPTER XI.

WE had not yet passed quite through the last settlements or suburbs to the Kansas city, when I perceived
that my beautiful young mare mule, which they had put
as one of the leaders to the baggage waggon, was so hot
and anxious that she drew the whole weight herself,
and that the Boh-hoy who drove her had not the re-
motest idea that when he cracked his whip and made
those execrable cries, so natural to Americans driving
mules, while he only urged the other three mules to
better action, he drove the fourth mad.   I spoke to
him of this several times, but neither he nor any one of
the other fellows seemed to care one farthing what I
said. They evidently considered that the waggons,

horses, and mules were, on the national equality system,
as much theirs as mine, and that I was but an appendage to be taken to the plains, when and how they
pleased. I saw at once that this democratically pleasing
notion on their part must speedily be put an end to, so I
said, " You, sir—don't you hear what I say? If you
want to hit your wheelers, do so quietly; but I won't
have any noise, and I won't speak twice in vain." The
Boh-boy stared at this announcement as if propounded
to one of the majestic people in a very unwonted manner; but he refrained from further noise, and so far, and
perhaps for the first time in his life, obeyed a master.
The fellow Mr Canterall then informed me that he was
very ill, and that he feared the fever and ague were on
him—a circumstance he knew very well before, but took
care to conceal till we had commenced our journey. On
this I administered to him from the medicine chest that
my friend Mr Buckland had prepared in England, and
told him to ask me for anything that he thought would
tend to his better health and comfort. The announcement he thus made to me, however, was very annoying,
as to him I had been told to look for the vigilance of an
honest and an industrious guide, whereas a fellow who
destroyed the effect of medicine by choking himself with
tobacco, and who was intermittingly prostrated by fever
and ague, and lying on his back in a waggon or on
the ground, retching for twelve hours out of the twenty-
four, was more of a burthen to me than of any useful
service. In this, the first commencement of my travels
in the desert, I had also to desire that the Boh-hoys,
who took turn about at riding and leading my horses,
should confine themselves behind my waggons and to
the line of march, and not go trotting on to call at

cabins in which their friends resided. Non-obedience
to these orders also drew down on them some emphatic
cautions, and thus, in the earliest possible instance,
although I had not then seen Capt. Marcy's most useful
work, "The Prairie Traveller," I endeavoured to impress on all my men that there ought to be, and, with
me, *could only be*, one head and one commander to the
little force of which my small party consisted.

After some miles of travel, all the dips in the track we
followed having small sluggish rills of water in them,
and on either side several yards of slough, through
which the waggons had to be driven with considerable
caution, the plains became very lonely, swelling into
moderate undulations, and in their brown state of
ripened grass, with only a flower here and there to be
seen, consisting of the snake-root and the wild sun-flower,
resembling in outline the petrifaction of a Dead Sea.

Having gone beyond the distance to which the sportsmen of Kansas city might have been expected to reach,
I gave orders for my setter, Chance, and the retriever,
Brutus, to be let loose, and mounting my pony, with one
of the old John Manton guns in hand, I gave Chance the
office to range the prairies in line parallel with the track
pursued by the waggons, while Brutus followed at the
pony's heels. Nothing could be more beautiful than the
way the old setter ranged these gently undulating plains,
delighting in the easy ground, and caring nothing for the
long grass nearly up to his back, for the grass of the
prairies is something like the English spear-grass, and
in no way, save when in winter there have been water-
courses to make it high and rank, offering any impediment to a dog; and, in addition to this, there are neither
stones nor thorns to lame him. While Chance was thus

ranging, I was occasionally turning my head from him to mark the course of my waggons, or to observe the interminable swell of the plains around me, when I heard the dear companion at my heels, Brutus, make those snaps with his jaws that I had so often heard from him when my dogs were pointing game in England, and I knew that he observed that Chance was at a stand. Turning my head, then, in the direction of the setter's range, for a few seconds no moving object was to be seen, but at last a stationary red spot, gone down very low in the grass, showed me that Chance was almost setting at something close under his nose. The point (for the stop to game which Chance makes ranges between a point and a set) looked like one made at a rabbit, for the English rabbit is dotted about all over the prairies,* though not even in the woods or creeks in any quantities, for which latter fact I can in no way account. On arriving within a little distance of the dog I dismounted, and, throwing the rein over Charlie's head, walked up to him, and asked him a question, the intonation and import of which he well knew, "Of what is it, old dog?" Chance's bright yellow eyes slightly turned towards me, and then resumed their fixedness; but I also observed that the impression of his face was unlike that which it wore on an assured fact of game under his nose; so I asked Brutus the question I had asked Chance, when, on stepping to the front and applying his nose, as if expecting to pick up a wounded bird or a rabbit from its form, he also made a very funny face, and I began to have wild notions of a poisonous snake. On this I peered into the grass, and, seeing a mottled surface un-

* It has since been denied by some people pretending to a knowledge of the matter, that the English rabbit is one of the wild denizens of the plains. The fact, however, is as I state it; the common rabbit of England is very frequently to be found in America.

der it, I put to it the muzzle of my gun and slightly lifted a land-turtle, rather larger than the palm of my hand, and which Chance had been pointing. Chance made a quiet but eminently disgusted face at the reptile, when, inserting the toe of my boot under it, I kicked the turtle a yard high, and Brutus dashed at it, but on nearly touching it with his nose he jumped back his own length into the air, grinning and shaking his head as dogs may have been seen to do at toads, and as if he had known that he was in a strange land of venomous reptiles, and that the innocuous little turtle might be one of them. "Leave it, dear old men," was then the word to my dogs; and I mounted my pony, and commenced, as it is called in America, my "hunt."

By this time, afar off in the prairies, I could see a few trees; and I knew that they marked the spot where, for the sake of fuel and water, we should encamp for the night: and right in my front there were some grass-cutters making loads of the dried small walking-sticks which in that country the settlers think it no libel to call "hay." Dear old Chance, having stood a few more land-turtles, and paused occasionally in uncertainty on the yellow-breasted meadow lark, he began to treat such refuse with contempt; then all at once, when at a long distance from me, I saw him with head erect wave to and fro, or undulate in his course, as the wind served him, and from his gallop come to a steady draw, and then to a decided point, with listless ears floating back in the wind, as if to catch my approaching footfall. Brutus had seen it too, for nothing ever escaped his notice, and had snapped his jaws in anxiety; so, directing Charlie to the spot, not in a hurried manner, for bad example to dogs is worse than the same to men, as in the first instance the better animal

sometimes suffers, and during my approach I saw Chance
gradually turn his head to be certain that I had seen him.
Arrived within fifty yards, I dismounted, and then the old
dog's ears, as well as the whole spirit within him, were
bent on the grass some twenty yards in front of him. Observing the direction of his eyes, I walked on and bade
him do the same, when up rose a lovely bird, very similar to the English grouse in flight and hue, but much
larger, and I fired, with only just time to know that the
bird had been struck, when I had to turn and fire on
others, that rose more to my right, one of which I killed
dead with the second barrel. While Chance sat down
delighted and anxiously watching Brutus, Brutus came
back with the first prairie grouse we had ever seen in
his mouth, and remained with it at my heels till I had
reloaded. This had happened in sight of my men, but
more so in the vicinity of the prairie hay-makers, one of
whom indicated with his hand the distant locality of the
first bird I had shot at, and which he had marked down.
Mounting my pony, and giving Chance the sign to bid
him range away so as to obtain the wind of the place indicated, I had the satisfaction to see him catch the taint-
ed air and soon after come to a dead point, when, on
reaching him, Brutus and myself found the first bird that
I had shot at, lying dead.

It was then becoming late in the day, and the grouse
were on their feed, and probably drawn off to some isolated corn-field around the nearest cabin. In my course
towards my waggons, however, which I could see had
halted near the trees before alluded to, I bagged in all
five prairie grouse, and on arriving at the head of Brush
Creek there I delivered them to the Boh-hoys, when,
having taken Chance to the little water-course among the

PRAIRIE GROUSE.

Printed by]

[Spottiswoode & Co.

bushes, of not very tempting water, I asked how far we had travelled, and received for answer, They were not sure, perhaps above eight miles, or it might be more. My use of the English setter this day proved that the advice of some of my friends in England, as to the nature of the prairies being so severe that English, dogs would be unable to face the work before them, was about as much beside the mark as a good many more of their " notions."

How the sight of these prairies made me long for English thorough-bred horses and foxhounds, and more so now, for I found myself encamped by the side of a small covert of seven or eight acres or less of beautiful lying, with no other wood within twenty miles of me, and nothing to stop a hound and horse but pace, and nothing to run over but grass, without foil of any description ; foxes certainly, and perhaps a small prairie wolf, not very much larger than a fox, is to be found in this beautiful situation.   Oh ! how I did indeed long for my old pack of foxhounds, for my favourite horse Jack o'Lantern, and for the presence of my brother sportsmen of old, Lord Cardigan, Lord Clanricarde, Lord Rokeby, Payton, Standon, Parker, Magniac, and Harry Boulton ; and *how* Charles Tolle-mache on his brown horse Radical *would* have galloped, for there was not a fence to shirk, nor a choice of ground to have induced in him a moment's pause.   If the American railway companies would only concede fair accommodation for the hounds and dogs of English sportsmen in vans, and permit them, if four or five gentlemen were in company, to have a " state carriage " to themselves, it would be perfect.   Do not look surprised, O democratic reader, but appended to a train on my return home I

12

found a very agreeable American gentleman in a beautifully furnished "state carriage" or saloon all to himself,
into which I was admitted by invitation.   As such things,
then, *are in existence,* why can they not be had for money ?
In the instance to which I refer, the state carriage assigned
for the journey to this gentleman was an act of courtesy,
because he was travelling to pronounce on, and be witness
of, if necessary, some facts regarding the interests of the
company.   To be able to take a pack of foxhounds to the
plains, and to carry with you their food and the means for
cooking (for to place them on the prairies behind a wolf
or fox without their being in tip-top condition would mar
the whole project), would of course cost a considerable sum
of money.   Thank Heaven, however, we have sportsmen
in England who could afford to embark in such a project,
and it would be well worth the voyage across the Atlantic
to be there to share in the splendid recreation the fulfil-
ment of such a design would afford.   It would be necessary to hunt from camp, and to repair to the plains amply
provided to sustain the condition and health of all.   Between the days of hunting there would be amusement
enough at the smaller game, and, supposing the English
hounds were to be sold when the experiment was over, if
divided into small lots they would realize a very large sum,
for in Kentucky and other states the English foxhound is
coveted beyond measure.

As to horses, for these America might well be trusted,
and steeds, at a ridiculous price as contrasted with value
in England, could be procured well able to carry 14st. to
any hounds in the world.   My chestnut horse, Taymouth,
was one of these, and happy indeed should I have been to
have been able to set his fiery spirit at the tail of

English foxhounds, flying over the easy undulations of the grassy and unlimited plains.

Arousing myself from such contemplations as these just narrated, I then, for the first time in my life, had to inspect the preparations for the encampment for the night, to learn, in case of hostile or thievish difficulties, the weak points of my position, and to endeavour to remedy them by the best means within my power. Then, indeed, I learned that which I have since read in Captain Marcy's book, how small a force mine was, supposing every man to do his duty, if in presence of hostile Indians, or thieves of any sort, to protect a camp from nocturnal invasion. My three waggons could not make a " crall," or fence, around my mules and horses, nor afford any protection to my party, while, at the same time, in order to obtain grass enough for grazing throughout the night, spots had to be selected for the various animals at such distance from each other that the tenant of the one picket-rope could not cross nor get entangled with the picket-rope of the other, nor reach the feed of corn in ear which was set down to each as the night's allowance. I at once saw that unless the sentinel on duty kept the most vigilant watch, any thief or thieves could crawl through the grass, and steal to and take up the furthest picket-pins, and lead off the animals that had been so tethered.

The first order given was to hang a mule-bell around my docile little pony's neck, and to tether him as much in the centre of the rest of the animal pickets as their position would allow. The next was, to be sure and drive the picket-pins most firmly into the ground, to husband space as much as possible, and to keep the camp to leeward of the animals, so as to have the chance

of hearing all noises, at least so long as the wind remained in a quarter available for that purpose. There
was a double advantage in this—not only that of hearing, but if by any accident the grass caught fire, the
wind would trend the flames away from the stock, instead
of in their direction. On the night of which I now speak,
my ambulance, as well as my dog-waggon, which I ordered always to be near it, was drawn up too close to the
baggage waggon, cooking-fire, and tent of my men, and
for the future I resolved that this mistake should be rec-
tified. Having determined to make the driving-seat in
my ambulance my place for dinner, I began to set out a
few biscuits and other little luxuries, and to think of a can
of preserved meats; but, on my asking if there was any
boiling water to heat the can, the cook, Wallace, informed me " that it was unnecessary, for the grouse were in
plenty, and cooked for immediate use." I confess, on
hearing this annunciation, I very nearly was angry, for
the feelings of the Old Country were not yet sufficiently
stilled, and I had thoughts of hanging the birds for
several days to make them tender; but the idea of where
to hang them, on the waggons, in that hot sun, and how
natural it was for my men to prefer fresh meat to salt
bacon, soothed me, and, having drawn the cork of a bottle
of brandy, I began to sigh for dinner. By this time
there could be no impediment to George reaching me
with the viands expected, although there might be to
others, for, according to orders, he had chained Brutus
and Druid to the two front wheels, and Chance and
Bar to the hinder ones, their chains permitting them
to push up the walls of the tent beneath the ambulance, and to sit either outside or in it, as they pleased,
and they resolutely refused to let the fellow called a

guide or any of his Boh-hoys approach me. George then waited on me, and very soon the half of a prairie grouse, very well dressed, and some broiled bacon, with a potatoe, were set before me, flanked by some excellently baked and lightly raised soda rolls; and I made an excellent dinner.

After dinner, at dusk, I took my gun and Brutus, and walked round the little covert for the chance of a rabbit, hare, or any beast that the approach of night might put on foot; but ere I had gone 300 yards, twilight, of which in those latitudes there is very little, had closed in, and the horizon became intensely black, though illumined every instant with lightning; the little wind there had been then fell, and I knew a storm was coming. Closely attended by Brutus, who seemed to think that additional care and vigilance on his part were necessary, I returned to my camp, and having chained Brutus up immediately beneath me, I took out my note-book to jot down the incidents of the day's march, and called for a lantern. The lantern, like everything else the care of which had been entrusted to Mr Canterall, was forgotten, so a candle did its duty, stuck up in a piece of old tin. This, however, at once, on that still eve of an approaching thunder-storm, and on that hot night, became so attractive to moths— moths which in size might have been the great-great-grand-mothers of any I had seen in England—that to write in comfort or keep the candle in was out of the question, so I put out the light and prepared for bed. "Bed!" Oh! how I longed for one, and how much I regretted that among the wonderful tales told me by English and American friends, who pretended to know all about the necessities of prairie life, not one had suggested to me an officer's tent and a camp-bed with sheets and pillow, nor even to take with me a mattress. Alas, I was about to lie down to rest on the uneven and hard cushions of the ambulance seats, with a double blanket spread over them, my coverlet a blanket, my pillow a gun-case; and but for health and strength, a frugal dinner, and plenty of exercise, I might have passed a most uncomfortable night.

Before lying down I served out to George a double rifle and a revolver, telling him to keep the young retriever Alice chained under his dog-waggon, in which vehicle he himself slept; when, hanging my loaded gun close at my side on hooks made within my ambulance for that purpose, with a waterproof cover on to protect it from the dew, and putting my revolver and long hunting knife in the corner close to my head, I sought to sleep through the first night on the plains. To me the situation was lonely and beautiful, and the constant tinkle of Charlie's bell invited rather than put to flight the visits of the drowsy god, and, amidst the low but increasing mutterings of thunder, and the very vivid and continuous flashings of lightning, I fell asleep, till a reverberating crash, of what seemed to me to be triple thunder, directly over my head, made me start, and then by torrents of rain I knew we were in a sort of deluge. It did not last long, however, and, as the storm decreased, the sound of my men's voices died away with it, and I fell fast asleep, and slept for a considerable time.

Things remained thus till past midnight, as far as I could judge, when I was again aroused by the cries of my men in confusion, laughter and oaths being strangely mingled, and I listened for further information. The cause of this "hullabaloo" was, that, while they were lying on their blankets, one of them heard, or fancied he heard, in their very midst, the rattle of the deadly snake, when, giving an alarm, they all jumped up together, and rushed out of their tent. A light being procured, they then searched the tent, taking up their blankets; but either the snake had fled, or never had been among them, and the camp was again very soon hushed in the arms of sleep.

My men awoke me next morning, as they prepared the fire for cooking breakfasts, led the horses and mules to water, and busied themselves in their listless, idle way about other matters, when, on looking out, I saw that the morning was a lovely one, all things spangled with the glittering drops of the night's shower, and ere I again retired, dear old Brutus put his paws up on the fore wheel to assure himself that all was right, and then, with a shake, scratched for himself a bed in the long wet grass outside the wall of the tent, to bask in the first rays of the rising sun. Having had my bath of cold water in the midst of the wet grass, dressed myself, and done justice to some coffee, bread, butter, and eggs, I called out to my men to " look alive," a thing they never did unless they were eating or swearing, and to " bring up the mules and harness them for a start." I then turned my attention to the rolling up of my own blankets, and setting the inside of my ambulance in order. While occupied in this, I had full opportunity of seeing the brutal way in which all these Boh-hoys used the animals intrusted to their charge, with the exception of Martin, who was certainly one of the best-conducted and most steady men I had. He was not only a quiet, well-mannered man, but I am sure he was a good rifle-shot, for when in the woods, and he could be spared, he used to kill squirrels and bring them into camp, always hit by his bullet in the right place. In the whole course of my journey through

the desert I had never cause to give Martin an angry
word.   The man who drove my ambulance was a quiet
man, and apparently a very good-tempered one; and
here my praise of those who had anything to do with
the mules and horses must cease.

Everything being in readiness, we made an early
start, and again mounted on my pony, with Brutus at
his heels; Chance ranged the prairies, and again made as
fine points as it was possible to see.   When we neared
any settlements, I returned to the line of march, and
gave my dogs, Brutus and Chance, a rest in their waggon, and, while they rested, Druid, Bar, and Alice
had their turn of exercise, so that they were all kept in
the best condition my means afforded.   Their food,
biscuit, which I brought with me for their consumption,
was very good, but a great ingredient to a working dog's
condition was wanting, and that was flesh.   Were I to
go to the plains again, if it could not be procured in New
York, I should bring with me from England the thing
we call " greaves," for though that commodity does not
answer the purpose so well as meat, still it is the best
substitute for it that can be obtained.

In the two days' march my observation showed me
that the prairie grouse were governed by much the same
rule as other feathered game.   Thus at daybreak, and at
that time of the year, they packed and sought the cornfields, when there were any, for their food, where they
were as constantly attended by every species of hawk
that haunted the plains, the common buzzard being
among them, as well as the hen-harrier and English
sparrow-hawk.   While packed and on their feed they
were very wild, but when their feed was done they
fled as far as possible from the location of the settler;

and I found them during the first part of the day dusting in the beaten tracks. Towards mid-day they were off to unfrequented places, and in shady covert afforded by rank grass and bushes.

On the march I again bagged five grouse, a brace of what the Americans call quail, but which are really their partridge, and a spotted water-rail. The meat on the prairie grouse is brown, but that on the wood grouse, which they call their pheasant, as well as on the quail, or partridge, is white, like that of the birds of the same name in England, — the pheasant or wood-grouse the whitest of the two. Of these beautiful birds of game the prairie grouse is the largest, the wood grouse the next in size : the former, if anything, larger than our black-game, and the latter about the size of, or rather less than, our grouse. The partridge or quail is larger than the quails we have in England, but less than our partridge, and they haunt the woods, particularly when they are adjacent to cultivated land. They lie well, do not fly far, and call together very soon, much as our partridge do, and they rise like them, but fly faster. They are in method of flight, gregarious habits, and haunts, precisely what a partridge in England is, with the one sole difference, that they, even in the earliest instances of the young brood, always hang about the skirts of cover. They have, however, one peculiar quality which the Californian quail has, but which our English partridge has not, and that is, they will fly from a dog or predatory animal for safety into trees. During my journey I had opportunities of seeing this, and have killed a brace of these beautiful birds from the tree above, with my double gun, while Chance was still pointing at others below ; those in the tree having taken him, as I have seen black game, pheasants, and snipes do, in the English New Forest, for a fox or some beast of prey, my habit of never speaking to or whistling to him not putting the presence of man into their heads. The little spotted rail, which I killed off Charlie's back, and of whom, during my journey, I could have killed many, was scarcely so large as the common English water-rail, but flew in the same manner; and generally where I found them, but not invariably so, there was some little swampy place or water not far off. On arriving at Cedar Creek, a distance of about twenty-five miles, so my men computed it, we camped on Monday evening on a running stream, and by the side of a covert, which might have held wolves, foxes, and deer, but of the two former only I saw traces.

Here then, for the first time, I saw quantities of that beautiful bird, the blue jay, of which I killed several. Hawks, of which I also shot several, and owls, the same horned-owls we have in England, frequently rose from the prairie grass, and one or two of these I also killed. The American crow, of course, was to be heard and seen at break of day, but they were always very shy. The snow-birds, in their summer plumage, and birds very like our larks, were also in plenty on the plains, and an occasional blue robin was to be seen in the woods, with many spotted little woodpeckers, and other small as well as larger birds, among them the lovely kingfisher.

On reaching Cedar Creek I ordered the five grouse and the rail to be dressed for camp dinner, but the brace of partridges to be retained for my breakfast on the following morning. The day, like the former one, had been very hot, and, from my being in the sun, I began to feel a feverish sensation in my mouth; but I was not

ill, nor did I fail in appetite, nor in sleep—rude, hard,
and rough as my bed within the ambulance was. Some
of my men, however, as well as George, began to complain; and, as far as that useless appendage to my camp,
Mr Canterall, went, he was retching, chattering in his
teeth, and ill more or less all day. He had by this time
donned a sort of leather dress, covered with fringe of the
same material, which he informed me was his " Indian
dress," but over this he put a sky-blue coat, of the oldest
English cut, with a cape to it, in which he looked excessively like a worn-out French soldier. As I proceeded
through the plains I found that this garment (no doubt
filched by volunteers from the Government stores, for I
found it was the greatcoat served out to that class of
warriors) was in fact the coverlet of all settlers.

If I remember rightly, it was on this day that I had
for dinner grouse-soup, thickened slightly with flour,
and flavoured a little with slips of bacon; and never in
all my life had I tasted of a more delicious decoction,
sent up boiling hot. It was so good that I called to Mr
Canterall to express my unqualified approbation, and to
tell him to compliment Wallace, who acted as cook; but
he told me he had made the soup himself, and that it
was nothing to what he could do when we got among
buffalo, antelope, or deer. Whether this was false or not
I am uncertain; but as I afterwards found that he could
not even stumble on the truth by accident, very likely
he assumed to himself the virtues of another. Here, with
this excellent soup, my present stock of potatoes having
vanished, a little bottle of milk, made to keep better by
having some brandy put to it, also became exhausted,
and the bit of fresh butter was all gone. With a caution
to Mr Canterall—but which caution, if it entailed on

him the progression of a hundred yards, he generally neglected — at all settlements, if possible, to purchase milk, butter, and eggs, I began to look up my hermetically sealed cans, and cans of excellent sardines, presented to me by mine host of the Planter's House at Kansas city, and to inquire for one of my bottles of sherry. With the parched sensation of fever in my mouth I felt sure that brandy was unwise, and, had I smoked or chewed tobacco, that those things should have been left off as if they had been poison.

On fixing the encampment on this the Monday night, I took care to have my ambulance out of earshot of the tent and waggon of the men, in order to avoid hearing their objectionable conversation; gave orders for the better protecting of horses and mules; and when they had their corn, which we had procured at a neighbouring settlement, I went my rounds to see that all was in better order. Mr Canterall, when not lying on the ground retching, always asked me "to leave it all to him," but never attended to any orders I issued after my back was turned. Monday night was very fine, and I went to bed and soon fell fast asleep.

CHAPTER XII.

On Tuesday morning the day broke without a cloud to
thwart the effulgence of the sun, and as the first ray
gleamed on the cold and heavy dew on the long grass
around my waggon, I was deluging myself on the ground
with water from the adjacent creek, and luxuriating in
the rub of a rough towel, preparatory to a breakfast on
the delicious partridge for which I had given orders.

A lovely sight in those still wilds is the rising sun,
rising at once above the undulating plains, as if over a
dormant sea! Strange cries of birds reached my ear;
and flocks of the blue jay, the morning rays glancing
from their splendid plumage, flew high and slowly over
my head, wending their way to the corn-field of some

settler, if there was one, or to other places where insect food abounded. During my vigils in the wooded creeks I observed that if there was a patch of corn to be found within morning's flight, there were sure to congregate a number of blue jays, in all probability not stealing from the crop, but doing a service, in such instances, and in England in other birds, too often misappreciated, by devouring the insidious insects that were unseen by man, and which but for the birds would have worked him ill. When the blue jays are thus seeking or returning from their feeding-places, they fly very steadily, and often very high, offering the prettiest overhead shots imaginable. In some of the States there is a penalty attached to those who kill them, arising from the fact of the general destruction of the feathered creation around cultivated lands, through the unlimited use of the gun, and the consequent increase of noxious insects. This protecting penalty is in this instance, however, in my opinion, an error, for though the blue jay no doubt destroys his quota of injurious insects, still, by the nature of the jay and the predatory habits of his class, by killing the young of smaller birds and sucking their eggs he ought to be regarded as a creature whose numbers had better be kept under, than protected for the purposes of further incubation.

The only way in which I can account for the general presence of the English rabbit, and yet their paucity of numbers, throughout the plains and woods wherein I travelled, is by supposing that they have so many enemies in the shape of wolves, foxes, skunks, coons, snakes, &c., that they are destroyed to an extent above their natural fecundity; for unless this is so, a soil so easily burrowed into, with the natural clefts in rocks,

and the steep and dry banks of the creeks, and a general
supply of food, must harbour and rear them to an extent
equal to that which, wherever they are left to nature, exists
in England. In the Cockney phraseology of America,
and on my first arrival, I heard of two sorts of rabbits,
one of which was distinguished from the other by being
called the " donkey rabbit," from the greater length of
its ears; the donkey rabbit being nothing more nor less
than our English and Scottish hare, their colour varying
in the United States according to mountain or plain, and
the effects of climate.

On this, the third day of my journey across the plains,
the heat of the sun was intense — so hot that I could
scarce bear my hand on the top of my own hat, and my
instep in the iron stirrup felt as if it was scorching. I
believe it is from suffering by the latter inconvenience
that on the plains a wooden sort of clog-shaped stirrup,
likened by me, in foregoing chapters, to a small coal-scut-
tle, is preferred, from the sun having less effect upon it.
In spite of the sun, Chance ranged the prairies through
the refreshing aid of occasional pools of black stagnant
water; and it being so intensely hot, I looked out in all
directions for snakes, having given orders to my men, if
they saw a snake of any sort or kind, but particularly a
rattlesnake or copperhead, not to disturb him, but to call
for me, that I might have full opportunities for inspection
before the reptile's death. The day's journey lay through
a good many settlements, and, as a natural consequence,
the adjoining prairies were utterly destitute of game; for,
in addition to the fact of the grouse being waited for at
daybreak in the corn-fields by the frontier men when
they came packed for feed, they were also hunted in the
cooler portions of the day, and either destroyed or driven

away; so during this day's march I shot nothing but
an owl and a hawk—in short, I saw no game of any sort
whatever.

Exhausted with heat, feverish in myself, but yet in
no way unwell, we reached a wooded creek called Black
Jack, and encamped for the night, and it was here for
the first time that I had recourse to my tin cans for a
dinner; from an adjacent cabin, however, we procured
some nice new milk. While looking among these excellent inventions for prairie life, I came across a tin can
of hermetically-sealed cherries, and, greedily opening
it, found the fruit excellent, free from any spirituous
combination, and, with a parched and feverish sensation upon me, intensely grateful and refreshing. The
beef and vegetables were too rich and too hotly spiced
for the state in which men find themselves when travelling on the plains; so, having eaten a little, I gave
the rest to the men, advising them to mix it with
their soup — soup made of water, flour, and a few
slices of bacon. To that weak decoction the spiced contents of the can, I afterwards found, made an agreeable
addition. During the night two of my dogs got loose,
for which George received a well-deserved admonition.
According to the computation of Mr Canterall, we were
then distant from Kansas city about fifty miles.

On the morning of Wednesday, the 28th of September,
having roused my men early, we were breakfasted and
packed up ready to start before seven in the morning,
and a glorious morn it was. During the march I gave
my men permission (those not actually on duty) to carry
their guns for any game that might by accident come
near them, and they shot a turtle-dove and a meadow-
lark, which brought out a caution to them from me, that

in future they should shoot nothing that was not large enough as well as good enough to be worth the powder and shot. My bag of game that day, of which the men killed one or two, was six grouse and two quails or partridge, Chance continuing to work whenever called on as if he had been bred and born beneath the rays of the American sun. As we were proceeding through a very lonely and desert extent of prairie, in one of those dips or bottoms before alluded to, crossed by sluggish rills of water, I observed a waggon stuck fast in the slough, and that a countryman of the poor emigrant who, with his wife and children, was thus hopelessly situated, passed by on horseback without offering the least assistance. I joined my waggons immediately, before they came to the slough, when taking a warning from the mishap of those in distress, Mr Canterall and myself sounded the ground on either side, and found a place at which my waggons crossed in safety. When safe over, not one of my men taking any notice of the sufferers, I gave the word to "hold on," which means to halt, and desired every spare hand to assist the travellers out of the mud.

This having been done by lightening the load, most of which had been taken out by the emigrant, we pulled and pushed at the wheels, one of the hinder ones of which was sunk considerably above the axle. On the man thanking me for the timely assistance thus rendered, I replied, "No thanks are needed, my good fellow. I am an Englishman, and most happy to have been of use to you. That is the way the men of England and America should help each other out of the mud; were I stuck fast, you would do the same for me." "That I would," the man replied, apparently

13

deeply interested in my being of the "old country;" "but for you, yes, sir, we had been here all night, or may be been obliged to leave our traps behind us. Going to Pike's Peak, you are, no doubt, yes, sir?"

"No, I am not; I am out for a hunt," I rejoined.

"What!" he exclaimed, in still greater surprise; "come all this way for a hunt—may be for buffalo? Guess that *is* a notion, yes, sir, *surelie!*" He then wished me good luck, and kept muttering to himself, "From the old country—out for a hunt—wall, I guess he *is* come fur away!" When, leaving him to his considerations in this matter, I joined my waggons in their progression, and on coming to a slightly rising undulation could see around me for at least an area of twenty-five miles, without so much as a tree, bush, or living thing to vary the monotony of the russet garb of the silent and solitary plain. On reaching a creek called "Marrian," we encamped for the night. Mr Canterall was very ill; George also was ill, and the same with several of my men; but not one of them would take the strong doses of calomel I suggested. Slight pills they did take, but to effect any permanent good very large doses of calomel were the only remedies, succeeded by similar quantities of quinine.

On the morning of Thursday, the 29th of September, the weather was still bright and fine for the morning's start, but up to noon the heat was mitigated by a fine, fresh air, which enabled myself and Chance to range by the line of march for several hours. The bag of game this day amounted to sixteen grouse and a quail or partridge, and, to my intense amusement, I killed the very same species of dotterel that, in the preceding winter, had puzzled us all at home at Christchurch as to what sort

of dotterel it was; the specimen then killed being, I believe, still in the possession of Mr Hart. I killed several
of these excellent birds while in America, and there they
are commonly called the "kildeer plover," but what their
true ornithological name is I am not sure. On approaching our camping place for the night, called "110 Creek,"
we saw some partridge or quails and an English rabbit or
two, at which latter I tried to get a shot, but in vain; and
here I had to take two of my men considerably to task for
the following ill conduct. The creek or brook by which
we were encamped was not only very steep in its banks,
but at the places at which the water in it could be reached
it was very boggy, and the drinking place for the mules
and horses was at a little distance from the camp.

I was sauntering along with my gun and Brutus in
search of game, when I heard all sorts of impious oaths
and the sound of blows arising from the wood in which
ran the brook, when, on reaching the spot, I found two of
the Boh-hoys beating my mule Black Jack, whom I drove
in my ambulance with my favourite mare, because he
would not drink. I pretty soon stopped this proceeding,
and asked these two unmitigated Boh-hoys if a sound
thrashing would be likely to increase the appetite no
doubt they always had for liquor. I then ordered them,
mules and all, to camp, and shortly after followed, to enjoy a very good dinner on the game killed in the morning.
After dinner it became so sultry that on lying down in
my ambulance to sleep I could scarce bear even a single
blanket in the shape of coverlid, and there being no air to
waft it away, the heat of the noon-day's sun seemed to
pervade the roof, to cling to my "fixings," and to lie dormant in every article within my waggon. While in my
first sleep the growling of advancing thunder reached me,

and then such a storm of thunder, lightning, wind, rain, and hail broke upon my camp, that, but for the low and wooded situation on which we were located, I should have expected my horses and mules, as sometimes happens, to have been driven from their pickets and forced to have flown before it.   It lasted the greater part of the night, and such continuous, close, and dangerous lightning I had never seen; and then it was that I thoroughly understood the necessity of conductors for the electric fluid which were usually attached to cabins on the plains, that, but for the life they contain, seemed scarcely worthy of such care.

As the storm passed over us the temperature of the air changed, and became so cold that I was glad to pull on extra blankets and a buffalo robe to keep me warm, and on looking from my waggon at break of day I saw that on the tin of water near my dogs there was ice of the thickness of a shilling, and that the long grass was silvered by the hoar of a white frost.   It is these atmospheric changes in temperature, as well as decayed vegetation, which make the plains so trying to the constitution of men —men even born on or in the vicinity of the plains; and if, when thus acclimatised, they suffer, why, no wonder that an Englishman like myself felt that he must take pains to keep in health sufficient to enjoy the sport for which he had left his own home.

That I resisted and escaped ague, and suffered, in comparison with my men and the frontier settlers whom I met in my travels, so slightly from the pervading fever, I attribute, not only to a naturally good constitution, but to the constant use of a cold bath and entire abstinence from tobacco.   My Boh-hoys informed me that their tent was flooded in the night, and George in his dog-waggon also got thoroughly wet through.

As we were now gradually approaching the Far West, I called Mr Canterall to me to ask about the state of the arms, when in course of conversation he related to me the following tale—the truth of the tale, like all others told to me, I do not vouch for—but of the events of which *he said* he had, in his youth, been an eye-witness; no doubt, from the brutal cruelty perpetrated, if he had been there he was an active partisan. He told me that when on the plains with a strong party, they were attacked at night by Indians, who attempted to get up a stampede among their oxen, and to some extent succeeded. They, however, repulsed the attempt, and collected their beasts. In bringing their cattle back to the camp, their sentinel on duty observed in the darkness, slightly illumined by the embers of the watch-fire, that there appeared to be more on the back of one of the oxen than was natural, and something made him so very suspicious of an Indian seeking ingress to the pickets of the beasts in that am-bushed fashion, that he fired at the risk of killing the ox, and this occasioning some bustle, what with the confusion, darkness, and smoke of his rifle, he lost sight of the iden-tical ox, when, on an inspection of the beasts, all was found to be right, and nothing seen of an enemy. The com-panions of the sentinel jeered him on the matter, and it was set down that the man had been mistaken. The next morning, however, showed that the sentinel had been vigil-ant and had but done his duty, for there on the spot of the occurrence, in the long grass, lay an Indian, shot through the loins, the lower part of his figure paralysed, but the redskin was, in other respects, quite alive and sensible.

"What did you do with him?" I asked.

"Oh," replied Mr Canterall, "as he couldn't stand up, we carried him and set him against a tree, and he made just about pretty pastime afore breakfast for the boys, a-shooting at him till they done him to death."

What sort of an appetite, I wondered, had the cruel rascals for their breakfast; and, thinking over the atrocity of speaking of pastime so connected with the misery even of a savage, I dismissed Mr Canterall to his duty. I had now been associated with my men long enough to make some observations on their characters, and of course I liked some much better than others. Among those that best pleased me were Martin, to whom I have before referred; also the man who drove my ambulance, Frank Tomkins, Philip Smith and Wallace. There are one or two more of whom I shall have to speak as my narrative proceeds; but up to this time my estimation of character went no farther. The men were miserably armed, their guns generally in wretched repair, and I believe there were not two revolvers among them, if any. Mr Canterall himself brought nothing but a huge old lumbering rifle, as heavy and unmanageable as a small cannon, with which he shot very ill; his shooting was as bad as anything I ever saw, so that, had my camp been attacked, as far as my men went, the fire on the enemy could not have been very effective. Although, in the first instance, I endeavoured to dispel the innate suspicions I had of the worthless qualities of Mr Canterall as a guide and trustworthy man, and even wrote to some of my friends at St Louis to say I was satisfied with him, still every hour my dislike to the fellow increased. Though I could detect him at the time in no more than shirking all personal trouble, neglect of strict surveillance over the nocturnal picketing of horses and mules, and ignorance of the situation of creeks for camping at, still I set down in my own mind that I had been saddled with a fellow much less fit for his situation than were two or three of those over whom he had been placed to preside, as the sequel will show.

On Friday, the 30th of September, we started from 110 Creek, the ground cool and wet from the night's storm, and heavy for the waggons, when, with Chance, I added four more grouse, an English snipe, and a couple of dotterel, or kildeer plovers, to the larder, and for inspection shot the male hen-harrier, by way of satisfying myself as to the presence of a British bird of prey.

In the first few days of our march I had used one of my favourite old John Manton guns, the splendid shooting of which had been deeply interesting to my men, as to which gun they had made many inquiries of me in regard to the price, and where such a gun could be procured, for a solution to which questions they received a reference to the shop in Dover-street, London. The distance at which I killed the grouse and other birds, at times on and off my pony, was to them marvellous ; and my subsequent experience showed me that American sportsmen, generally, from the fear of missing, perhaps, while I was with them, or from habit, or the estimated range of their weapons, only attempted the shorter distance. The weather being so hot, and aware that it was a slightly less heavy gun, on the day to which I now refer, I had opened the case containing my new double shot-gun made for me by Mr Pape, of Newcastle-on-Tyne. On opening the case I never saw anything so neatly arranged and complete as the compartments within it, nor, on putting it together, a more superbly finished gun, stocked by the pattern of my favourite John Manton, and carrying nearly the same charge. My men were delighted with the inspection of it, and very soon had again to see the birds fall at distances they deemed impracticable.

With Pape's gun in my hand, still accompanied, by Chance and Brutus, I saw a pedestrian coming towards my waggons, who suddenly jumped many feet out of the road, while, at the same time, he as well as my men called and beckoned to each other. I also saw that he picked up a stone and flung it at something in the road. My men then, in obedience to orders, called to me, and, on arriving near enough to hear what they said, I heard the word "Rattlesnake!" So I called to George to secure my dogs, and walked up to the spot. Now, then, at last had arrived one of the longed-for moments in my travels in which I was to see alive, in his own wilderness, one of those deadly snakes, that I had hitherto met with but in a glass case in the Regent's-park gardens; and I can scarce describe the sort of charmed sensations that beset me! It seemed that between my waggons and the advancing pedestrian, in the hot dusty track, and sunning himself, lay a rattlesnake, so far in the middle of the beaten way that his retreat was impracticable as well as impeded to some extent by the wayfarer's stone, which had struck him slightly near the tail. Finding himself cut off from shelter, and thus assaulted, the reptile had coiled himself up for mischief, much as our adders or snakes do in similar situations, with his head overlaying his coils, and bent back in a threatening position and ready to strike at anything that came within reach of his poisonous fangs. Having snake-boots on, the reptile being not much over three feet in length or thereabouts, I went up close to him to examine his real appearance on his native prairie, and to watch his action. On advancing the muzzle of my gun close to him he instantly struck it with his teeth, looking, I thought, considerably disappointed that he could make no impression, when, on immediately inspecting the new and beautifully-browned barrels of Pape's shot-gun, I could not observe that he had left any trace of poison on the metal—not even the dampness of breath. I then put the gun to my shoulder and shot him in two, both pieces flying yards into the air, and eliciting an expression of approbation as to the hard shooting of the gun from some of the men. Mr Canterall then, by my direction, cut off the rattle, and the line of march again proceeded.

In all the opportunities I had for conversation with the natives, I of course picked up some strange tales in regard to rattlesnakes, and among them I was assured that, at times, immense numbers of these reptiles had been encountered on the plains, apparently in the act of collected migration through the grass ; that mules had been known to detect the presence of rattlesnakes thus congregated, and that in one instance, in an Indian camp, a number of snakes in the act of migrating had come in upon the horses and had bitten and killed some (to the tribe) valuable animals ; but, again, in no one instance could I hear that cows, oxen, or bisons had been known to die of the reptile venom. The day continuing intensely hot, and as our start had been an early one, and there was yet a long way to go, I halted the waggons at noon at 112 Creek, to give my men their dinner at their favourite hour, to refresh my mules, and, while this was going on, to amuse myself with Brutus in wandering along the wild banks of the almost stagnant water in the creek in search of sylvan adventure. There was no game, however, to be found ; as to wild turkeys and deer, said to haunt all the creeks on the route, their tracks, slots, or usings were not to be detected, so I ceased to expect them, for any man skilled in woodcraft never need see animal nor bird to be aware of their vici-

nity,—the signs of their haunts set him on the alert, and
warn him full surely of the game in store.

Having heard as I left my camp that my dog-waggon
wanted repair, I asked two frontier farmers, whom I found
collecting the rough stuff called hay, if there was a blacksmith among the few cabins where the beaten track
crossed the creek, and received a reply in the affirmative.
Having entered into conversation with them, as to the
scarcity of deer, turkey, and grouse, and told them whither I was going, they asked me if I had heard any news
from the direction of " Pawnee Fork " as I came along,
for they had heard a rumour, but nothing certain, that a
difficulty had arisen with the Indians there, and that white
men had been killed. Having told them that I had heard
nothing of the matter, I then returned to camp, and, on
inspecting my dog-waggon, found that the journey could
not be continued till it had been repaired, and that Mr Canterall was lying on the ground retching so violently that
I thought he must have broken a blood-vessel and died.
As usual, his disreputable nephew, who seemed, from his
unmitigated idleness and the invariably hideous blas-
phemy of his language, to have come into the world for
no other purpose than to be a nuisance to all connected
with him, was seated at his side by way of an excuse to
do nothing, so I bade that young gentleman get up and
make himself useful. One of my men having found the
blacksmith and brought him to inspect the dog-waggon,
the following amusing conversation occurred:

" Can you mend the spring of my waggon, which that
rascal of a coach-maker at Kansas city sent out in a damaged but disguised state, and make it fit for the journey
without delay ? "

" Guess I can."

" Well, then, I will send it to your forge directly, as I am in a great hurry to go on, and you can do the job out of hand, or ' right away,' as you call it."

" Guess there's no great hurry in sending the waggon to my location ; t'll do an hour hence."

" An hour hence ! why, my good man, I tell you I don't want to lose a moment ; so, if you can't do as I desire, I'll patch the waggon up myself till I come to some other place."

" Wall, guess then there's no harm done, if you find that ' some other place ; ' so good morning."

"Stay, my friend," I said, laughing, "you are all of you here about the rummest set of chaps I ever came across ; don't you like money ? "

" Guess I aint *hostile* to *that*, no how ; but I must have my dinner first."

" Dinner !" I rejoined ; " can't you mend my waggon, as it is of moment to me, and dine afterwards ? "

" Guess not, to-day."

" Why ? "

" Killed two _Perarie_ chicken yesterday, and got them for dinner just now, so must eat them first."

" Well," I cried, seeing that there was no work to be got while a democratic lord was hungry and in knowledge of the possession of something to eat, " my waggon shall be at your forge in an hour," and we parted very good friends. Taking my seat in my ambulance to eat a biscuit and to drink some weak brandy-and-water, I saw a little grassy mound beneath the feet of my men, and under the tree near the fire, on which they kicked and trampled as if it were but a mole-hill ; its length and shape, however, arresting my attention, I asked them what it was ?

" The grave of some stranger who died a few days ago
by the track side, so they buried him here," was the careless reply.

" Does anybody know who he was ? "

" Reckon they don't, governor ; there's another grave a
little further under the trees. Fever's strong in this
location."

" Well, then," I said, " as there is plenty of grass and
shade for all of us, without kicking holes in the turf which
they have had the decency neatly to lay over the poor fellow's head, just move the things a little further off, and,
as you may sleep in that way yourselves some fine day
(no one knows how soon), don't indecently disturb the
rest of death."

My orders being obeyed, I again went into the wood
round the creek, with Brutus, and met Martin coming out
with water. "Mind, sir," he said, "there are quantities
of snakes. I have just killed one, and in the bushes close
by I saw a rabbit." On returning with him to look at
the reptile, there lay as fine a specimen of the black snake,
not venomous, as could well be seen, which on measure-
ment proved to be more than five feet long. In the bushes,
also, Brutus found the rabbit, which a snap-short from
Pape's gun enabled him to bring me in his mouth ; it was
a fine, full-grown young one, of the English sort.

The waggon took some time to repair, but the Vulcan
of the plains made a good job of it, when, on preparing
to start, I found Mr Canterall so ill that it was impossible
for him to move ; so, there being by chance a travelling
medical practitioner at the cabins, and one of the inhabitants being known to Mr Canterall, I resolved to let this
fellow be put to bed in a hut and medically attended, while
I simply shifted my camping ground to better pasture and

a safer position on the other side the creek. I then encamped at a place called Logchain, on the banks of the
very steep-banked brook, where the water, by making a
bend, protected me on three sides.

While arranging our camp, and attending to the picketing of my horses, my eyes were greeted by the
appearance of an Indian (the first I had seen), completely equipped, on his pony leisurely, but I thought
oddly, coming towards us, when I soon saw that he
reeled in his saddle, and, on closer inspection, that he was
in such a state of real or pretended intoxication that he
could hardly maintain his position. Arrived at my
camp, he said something to the first of my men he met;
the only word I could make out was a demand for whiskey.
I called out " No ; " so my man, as I had previously forbidden either spirits or powder to be given them, gave
him a piece of tobacco, and made signs that he had
nothing else. The little redskin—for he was but a poor
specimen of a man—then stared at me, turned his pony
and rode off, setting at rest any suspicion in my mind as
to his being a spy, perhaps in a pretended state of intoxication, by reeling several times on the back of his
pony, and then, the girth slipping, tumbling, blankets,
paraphernalia, and all, heavily to the ground. The
poor docile pony stood still directly, and while we were
laughing the redskin picked himself up, re-arranged his
riding-gear on his pony's back, and rode slowly off into
the prairies, my dogs, who were luckily chained up,
perfectly furious to get at him.

The Indian ponies that I met with in my travels,
though high enough to be called of the cob size, did not
take after that particular shape ; they were low for
their length, and calculated for quick purposes rather

than for carrying or drawing heavy burthens, and none
that I saw could have borne me with any ease up to a
bison. I should, therefore, counsel all hunters seeking
the prairies, if they ride above 10st., not to listen to the
advice of obtaining horses out among the Indian tribes
for the purposes of hunting, but to select them from the
last frontier town ere they set foot upon the plains.
Tired with the exertions of the day, I went early to my
blankets, and slept till sunrise the next morning.

CHAPTER XIII.

ON sending down to the cabins on Sunday morning, the
2nd of October, I found that the travelling doctor,
whom I thought was a sensible man, had so handled Mr
Canterall that he would be well enough to march, and I
ordered a start accordingly. This was the eighth day of
travel on the plains, when, though I had tried all likely
ground in the line of our journey for deer and turkeys, as
far as opportunities would permit, I had seen none, nor
even their tracks, the daily bag consisting of grouse,
partridge, snipe, and dotterel or killdeer plover, rail and
rabbits. I had, however, seen both the common teal

and wild duck, and some plover of the larger kind, but
had not killed any of them.   Soon after striking camp,
old Chance brought me up to some prairie grouse, of
whom Brutus picked up three that fell to my gun, as
also a dotterel.

The day then becoming very hot, I joined the waggons,
and met a single horseman, who reported that the tale
heard by me at the last camping place was true, for that
two thousand Indians in war paint had plundered the
Government mail and murdered the three Americans in
charge of it.   Shortly after•this horseman had left us, a
considerable drove of oxen came from the same direction,
followed by two waggons.   The retreating whites looked as wild as their cattle, and all seemed to have been
pricking in hot haste out of danger's way.   The first
travellers in the first waggon confirmed the news of the
difficulty; but from those who were bringing up the rear
I gained some very interesting intelligence.   They informed me that they were some of Mr Peacock's men, a
frontier settler of considerable influence, whom I had
left at Kansas city, and to whose settlement I had promised to pay a visit in my way to Pawnee Fork, whence
he assured me I could reach the buffalo, while at the same
time I could avail myself of the safety of his property
and the protection of his men during the time of my
stay in that vicinity.   His protection and his hospitality
were thus alike out of the question, for his men had
abandoned his lands, and those with whom I thus spoke
advised me on no account to think of continuing my
hunting excursion to Pawnee Fork.   My informants
observing that I doubted the estimated force of the
Indians, and that I was inclined to think their accounts sa-
voured of a national inclination to magnify, when speak-

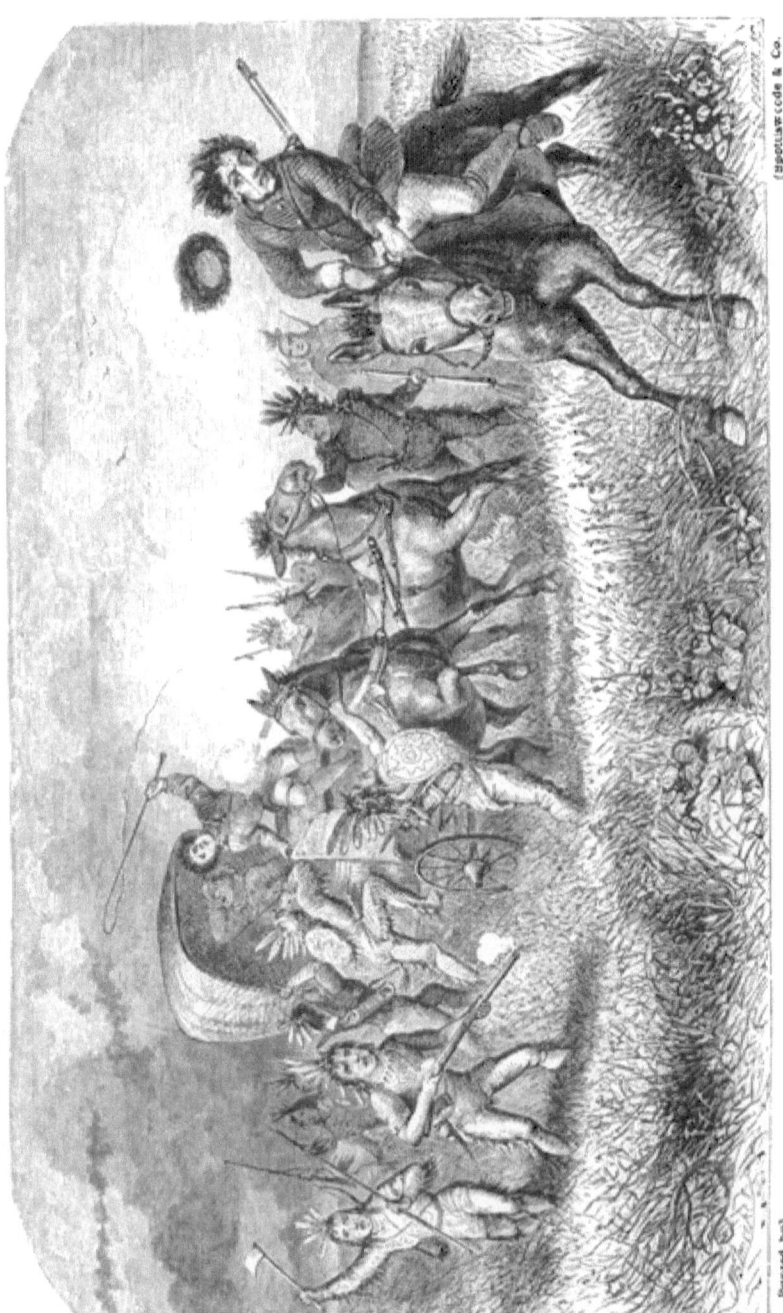

ATTACK ON THE GOVERNMENT MAIL.

ing of the attack on the Government mail, one of them exclaimed, " That he had some right to know all about it, as a brother or a brother-in-law of his (I forget which) was one of the two men who were murdered while in charge of the mail, and that the third man, who had fallen wounded into the long grass, but had contrived to escape with his scalp, had conveyed the news to the military station at Fort Riley." They all advised me not to proceed, and we then continued on our respective ways.

Thus, then, I had gained some important tidings which obviously demanded serious consideration, and during the rest of that day's journey I pondered well over the matter. To return whence I came was in no way a part of that consideration, for I had set out with a resolution to kill buffalo within a limited time, and, therefore, nothing that could be so accomplished should be abandoned. Mr Canterall having been present at the delivery of the news, I called him up to hear what he had to say, but to act, in regard to his opinions in the matter, with just as much attention as I thought they deserved. This fellow made light of the news, and advised my going on direct for Pawnee Fork; and when I asked him if, in giving this advice, he disbelieved the news, for if the news was true it was obvious folly to risk a difficulty with a large body of Indians with so small and ill-armed a force as mine, he said he did not suppose the news was all false, but we might, perhaps, escape the Indians, or perhaps, if we gave them what we had, they would not hurt us. Now that I know the man, my opinion is, that had we got into any difficulty, he would have been the first to desert me, and, if I had been over-

14

powered, to have shared in the theft of all the property
I had, making his own terms with my assailants irrespect-
ive of any interests of mine.

Having heard all Mr Canterall had to say, I resolved
to take him into my counsels no more, but to keep my
line of march as much to myself as possible, issuing
orders merely for the day. With these resolutions, I cast
Chance off again for game, and ere noon added three
prairie grouse and a dotterel to my larder. I saw a few
large plovers, and had a long shot at them, but without
success. The day being again intensely hot, we halted to
refresh the mules and to give my men their dinners at
noon at "Rock Creek," a running stream of larger dimensions than any of those on whose banks I had hitherto
encamped. We were busy with our outspanning arrangements, when one of the men, who had been to fetch some
water, came back to say that there were wildfowl in the
brook; so, taking Pape's shot-gun and Brutus, I crept
down the bank of the creek, and soon saw four blue-
winged teal seated pluming themselves on the brink of
a shallow part of the stream, when, having achieved an
easy distance for a shot, I killed a couple of them with the
first barrel, and one with the second; and Brutus was very
amusing in trying to get them all into his mouth at the
same time, so as to insure their capture; having one quite
dead in his mouth, he did not like to leave the other two,
who, shot in the head, were tumbling deliriously about
on the surface of the water. This he could not do, so he
brought them separately; and more plump birds I never
saw. They were larger than our common teal, and instead
of having the green or duck wing, that portion of the wing
was of a beautiful rich blue. Since killing these teal, I
have eaten (though I never shot one) the boasted canvass-

back duck, for which I had always heard that America was so famous, and I am perfectly convinced that there is a great mistake in claiming the superiority for the larger bird; the blue-winged teal is infinitely the best water-fowl I ever tasted in the whole course of my life.

Being so near the waggons I took my birds home, and then went up the river after the other teal. While in this pursuit, I came on a solitary cabin, situated so that the river and its steep bank protected it on one side, while on the other three there were rails infinitely stronger and more compact than were needed simply to fence in cattle. The place at once showed me that whoever built that "fixing" had defence from Indians in his mind, as well as the mere confinement of his beasts. Not being able to find the other teal, and it being intensely hot, in addition to my being beset by a very feverish sensation, I was returning home, when, wishing to get into the shallow stream to walk along it in my waterproof snake-boots, instead of having to force my way through cover, I went down the steep bank with a rush, and as I broke through the screen of wild vines and other creepers with a dash into the water, I most unexpectedly and abruptly almost ran against a tall, powerful Indian. He was apparently cooling his large, muscular, and shining-brown legs and thighs in the shallow water. As I thus came suddenly upon him, he was with his hands holding up his "breechclout" as high as he could to keep it dry, but, letting it go, he drew himself up to his full height when he saw me. He was the finest man of the Indian tribes that I had yet seen or thereafter saw, but a more ill-looking, ferocious scowl of surprise than he bestowed on me never disfigured even the face of a savage. We stared at each other for some moments, my opinion being that I had never beheld a much more powerful and certainly not a more ill-looking rascal, and his, perhaps, that he had never seen a better-armed man, for I had a double gun on my shoulder and a revolver and an immensely long East Indian dirk, or *couteau de chasse*, in my belt, while at the same time, close at my heels, with a furious desire to bite him which I could with difficulty suppress, was a large and faithful dog. At the back of the Indian was fastened what seemed to be his blankets, and, above the large sort of bundle they made, his bow and arrows, and these latter were in a case.

Having gazed at him to the full, and consequently christened him after the manner of " Satanka, or the Sitting Bear," of whom I shall have to speak anon, " Tackalyoucanget," or the " Big Thief," I marched off down the stream and continued looking for fowl, and on my returning to my camp found that, after seeing me, the Indian paid my men a visit, asking for whiskey, but getting nothing but tobacco. I also regretted to find that the huge " Kaw," for he belonged to that tribe, being alone, and my men being nine to one, in direct variance with my orders they had teased and jeered him, sending him away in high dudgeon; and had it been anywhere where he could have mustered his tribe, or had his miserable tribe been a warlike one, and in any strength, perhaps their jeers might have got us into trouble. As it was, the incident made me resolve on going my rounds at night, and seeing that all picket-pins were properly secured. Having refreshed the men and mules, we then put to, and regaining the beaten track we met more frontier men driven in, all of them agreeing that in the direction of Pawnee Fork there was at all events a large force of redskins in war paint—those who had attacked and murdered the Government mail men—and that it would be madness in me to attempt to hunt in that direction. They all advised my turning back; but, thanking them for their advice and good wishes, I gave no one any intimation of the course I intended to pursue; and in the afternoon, on reaching a place with wood and water called "Big John Spring," I encamped for the night, intending the following day to reach "Counsel Grove," where Penn had signed the treaty, and when there to glean all further information in my power.

On the following morning we passed Counsel Grove, and heard all the ill news confirmed, and were told that another mail was overdue, and supposed to have been stopped by the Indians. In addition to this intelligence, I was informed that some buffalo hunters from New Mexico, whom I had previously heard of as being out on the plains, had been attacked, though a strong party, and had all their mules taken; so, quite according to the old adage, that ill news travels fast, I had plenty of reasons briefly afforded me for taking measures to avoid a collision, against which I was but inefficiently prepared.

At Counsel Grove, then, very aptly I communed with myself, and resolved on my future course, and at once called up Mr Canterall.

" Well," I said, " what do you now advise, with all these difficulties before us ?"

" Oh, wall, just to go on to Pawnee Fork—that's a fact."

" Do you ?" I replied; " which is the way to Fort Riley,—do you know it ?"

" Fort Riley !" replied the fellow, with very evident dislike; " it lies away here to the north; oh, yes, in

course I know the way, but it's out of our line, gov'nor—
that's a fact."

" It does not happen to be in any way out of my line,
so if you know the way set the mules' heads for it at
once," I replied, " and no *two* ways about it ;" and Mr
Canterall then went to inquire the route, of which I saw
that, as usual, he was profoundly ignorant.

The fever which had been trying to lay hold of me
began then to make itself felt; great lassitude assailed me,
and my gun became heavy to my hand, so much so that
I confined myself, after the cool of the morning had passed, to my ambulance waggon, which, when I was in it, I
generally drove myself. While on this portion of my
journey, and when occupied with Chance picking up a
few grouse, in one of the boggy dells in the track which
so frequently occur on the plains, I again saw an emigrant
waggon stuck fast, and while several men with horses and
mules were endeavouring to rescue their goods from that
position, from the distance I beheld my men cross the bog
a little on one side of them, and continue on their way,
without any further notice whatever, but that of having
made the difficulty of their countrymen serve to guide them
from a similar danger. My waggons were so far ahead of
me that in the state of lassitude in which I was I could not
attempt to overtake and stop them, but on reaching the
party in the slough I told them that I deeply regretted the
incivility of my men, and that had I been there every
mule and man I had should have assisted in the liberation.
Among the people was one who seemed more intelligent
than the rest, when, having surveyed me a moment, he
said, " Are you the English lord that's come a hunting
to our plains,—Lord Berkeley, as I read on in the papers?"

I replied, " My name is Berkeley, and I am from

England, and out on a hunt;" on this we shook hands, and he wished me every success. He also confirmed the news as to the outbreak of the Indians, and told me that he had left Kansas city, as I understood him, shortly before my arrival at St Louis, but that he had returned on account of the ill-feeling among the redskins, which was spreading through all the tribes, who were more or less assuming a threatening attitude.

The weather continuing intensely hot, I again halted at noon to give the men their dinner and refresh the mules at a place the name of which sounded like Neocio, Nunchio, or Nocio, about ten miles from Counsel Grove, towards Fort Riley; but though I wrote it down at the time the letters have become effaced. We halted near a little rill of water, but not where any fuel could be obtained; my men, however, had provided themselves in some unaccountable way with wood to boil their kettle, and were in the midst of cooking interests when two frontier settlers rode up, one of whom thus and rather abruptly addressed me:

" D'you let your men steal people's property as they go along?"

" Steal!" I replied, " not if I know it."

" Wall, I reckon, then, they *have* stolen my rails, and I'll swear there's some on 'em on that fire."

On this I called to Mr Canterall, and said, " How is this? My men are charged with having stolen the wood they are using?"

" Well," said Mr Canterall, " I jest told the boys they might pick up a bit of dead wood as they passed a settlement, that's a fact."

" Yes," rejoined the frontier man, " and that piece of dead wood, as you call it, I guess was one of my rails. I can't keep any of my fixings, no how, while such like as you, emigrants and what not, keep a-going along these tracks, and I've been no end o' times a looking out to catch you."

On this my men joined in rather rudely ridiculing the alleged value of the rail, and inclined to give the two frontier men the rough side of their tongues. I bade them be silent, and, getting off my ambulance seat, looked at the wood they had, and saw that there *was a possibility* of calling it a rail. On this I asked its value, a mere trifle, and at once paid the sum demanded, giving my men a caution, that if they took anything from any one for the future, I should stop the value of it from their wages. Altercation, numbers against one, again arose, but was as speedily repressed by me, and, on the recipient of the money thanking me, and taking his departure, I told him that if ever anything was improperly done by men in the service of an English gentleman, I begged him to recollect, and to tell all his countrymen so, that we held it in the old country to be our duty, under all circumstances, to afford redress, and to be answerable for the conduct of any men in our employ. The owner of the rail alone rode away, his friend and companion, with an increased friendliness of manner, accompanied me to my ambulance, and remained for some time in conversation. He seemed much pleased with what he had observed, and greatly taken up with the distance I had come, and the complete fixings and arms in my belt and within my waggon, as well as with my dogs. In going away he most heartily wished me success, and added, "Your track, yes, sir, passes just afore my house, and if you'll let one of the boys come to my door I shall be very pleased to place some nice

water melons at your service." As we passed, my friend
of the frontiers was there with so handsome a gift of
melons, that there were enough for me and for my men,
and to last me several days.

Suffering considerably from fever, but without ague,
I travelled on till dark, and then encamped at some place
the name of which I have forgotten, and then again considered the step I had taken, and resolved that the course
I was pursuing was in fact the only safe one left to my
acceptance. My position was precisely this. In front
of me, towards Pawnee Fork, was, if over-rated, still a
considerable force of hostile Indians—Satankee, or the
Sitting Bear, a noted warrior, at their head ; while at the
same time, immediately around Counsel Grove, there
was on the following day to be a gathering of Indians to
receive Government money as the price of their lands,
and to get drunk with the consequent and unlimited
possession of whiskey. It would not be wise, then, to
delay, nor to subject myself, by waiting where I was, to
the chance of quarrel with redskins in inebriated fury.
In addition to these two agreeable facts, there was another, and that was an expected fiat from Government as
to whether the settlers, self-located on the plains near
Counsel Grove, were or were not in legal possession, or
whether they were but trespassers on Indian bounds.
Here then, in regard to the expected fiat, was placed before me the weakness and inability of the Government of
the United States to enforce the laws of the country ;
and I arrived at the knowledge of it, thus :—From time
to time, at my camping-places, the frontier settlers would
come, in their straightforward but invariably civil way,
to my ambulance waggon, to converse with me on my
journey and look at my arms. They were never rudely

intrusive, but, on the contrary, if homely, still excellently-conducted men; and we were always the best of friends. One specimen of this style of men, as well as the state of the affairs around them, will suffice; and I will give the reader a sample of our conversation, in the phraseology common between us.

"Well," I said to a frontier man seated on the pole of my ambulance, "you are a likely-looking chap to take your own part in a man-to-man difficulty; but what I can't understand is this. You are many of you taller than I am (indeed, I used to think myself tall till I got among you), but though you are twice as young as I am, you have lost the roundness of muscle, you have no colour, and your cheeks are hollow. What the devil has been the matter with you all?"

"Matter, sir, I reckon! Yes, sir, your cheeks would be hollow too if you were located on our plains. It's enough to kill a hoss."

"Why, I thought the plain was the healthiest spot in the world. In England we are taught to boast of your Indian summer."

"Boast! Guess you Englishmen know much about it! We get the fever and ague here, come every summer, and unless you break the head of it at first, guess it sticks to you pretty close. I've my shaking day, and so has most on us, as regular as you can fix it!"

"Well," I continued, "you had better keep yourselves in travelling trim, for just about now your Government is to decide whether you must turn out of your locations, and whether or not the land you are on belongs to the Indians. Nice thing for you if you've got to cut it!"

" Cut it! What's to make *us* cut, as you call it—yes, sir ! "

" Why, the decree of the Government."

" The Government ! guess *that* won't do it ! "

" Why, don't you mean to obey your own laws ? "

" Guess we *do*, so shall *stay*, and hold on *considerable tight !* "

" How do you mean? The Indians will come and turn you out, as they have a right to do ? "

" Wish they'd only *just about try it*, that's *what* we wants it *to* come to. Shoot them beggars *down* to *a* man then ; wants to get rid on 'em, but *no* such luck ; *wish* the Government *would* only give us the difficulty ; 'twould be as good as dollars ! "

" Well, I always knew you chaps went pretty strong ahead—all masters and no men. Suppose I came out on lands where the Government had decreed their sale at a dollar and half an acre—cheapish fixing that for some of the soil I have seen, and wished to buy ? "

" Well, you'd find us in no wise freendly."

" How so ? "

" Only jest as this ; if any soul alive came out to buy the land from under us, it's just the right of a free man to shoot the interloper down, and to keep what Heaven has given him. Reckon you'd find us each with his revolver, knife, and rifle, and riled for mischief right away."

" I don't doubt it, my good fellow, and I don't mean to · buy property nor share in the blessings and justice of your happy land ; shall we have a hunt together for a deer up the creek before I march to-morrow morning ? "

" To-morrow," said my friend, musing; " no, can't to-morrow, it's my shaking day," alluding to intermittent ague, " and I arn't fit for nothing."

" Well, God bless you, my boy; take an Englishman's advice, leave off chewing that mouthful of poisonous tobacco and spitting up the juices of your stomach to the sallow colouring of your cheeks; drink none of the stuff you call whiskey, and instead of spirits buy calomel and quinine, and grow some fruit and vegetables in your never half-cultivated garden, and then you'll look a jollier fellow."

: We always conversed in this familiar way, and a finer lot of fellows than the generality of the frontier settlers I do not wish to see; but, alas! go as far ahead as the American frontier man may, two things for ever keep pace with him—tobacco and whiskey, while religion and roads are left behind. Whiskey—bad whiskey, or even spirits of any kind—are the worst things he can drink with his liver in dangerous disorder, which is, in fact, the cause of the fever on the plains; but as whiskey affords him temporary warmth in the aguish chill, and enables him still further to debase the witless or thievishly crafty savage, whiskey is the deity of the cabin, and will continue to be the source of evil to the destroyer and the destroyed, or, in other words, to the white man and the redskin of the plains.

The cause of the Indian outbreak which lay in front of me was as follows, and I give its correct details, inasmuch as I subsequently made the acquaintance of the smart young officer (Lieut. Bayard, of the United States army), who, in his strict duty, had indirectly created the difficulty, and the narration of it cannot but be curiously interesting to my readers:

During the summer an army of Indians, reported in the *Dubuque Herald* as amounting to 2500 men, and consisting of the Kiowas and Comanches, had been collected in a threatening attitude in the vicinity of Pawnee Fork, Walnut Creek, and that portion of the plains.   They had, however, been watched and overawed by detachments of the United States army until immediately preceding the 26th of September, about which time, as far as I can learn, the troops usually begin to seek their winter quarters.   No sooner had the troops commenced to leave the plains, than on the evening of the 21st of September a party—consisting of Kiowa Indians, about fifteen in number, under their leader, a petty chief, called "Pawnee," who, though not belonging to the tribe of that name, was the brother of Tehorsen the chief of the Kiowa Indians, and son-in-law to Satanka, or the Sitting Bear, who is the war-chief of the same tribe—came to the trading post on the Santa Fe road, and with threats demanded liquor. This the white men at the post refused them, and barred the doors and windows of the cabin. The redskins mounted the roof, and, brandishing their tomahawks, threatened the inhabitants with death unless they immediately complied with their demands.   The whites within the house, however, under promises of future gifts and by persuasion, pacified the savages, and induced them to retire and leave their leader Pawnee still in a state of conference.   As soon as his followers were thus disposed of, they then seized the petty chief, disarmed him, and sent an express for military aid, when Lieut. Bayard, in a forced march of twenty-eight miles, speedily arrived with a detachment of his men, and, according to orders, received the prisoner into military custody for having threatened the life of an American citizen.   The detachment, when at the trading post, dismounted to refresh after their hurried march across the burning plains, when shortly after word was given to mount, and Mr Bayard ordered the prisoner to unfasten the larriet of his pony from the picket-pin, and to prepare to accompany the escort. The redskin had previously been made to deliver up his arms—at least all arms that were visible—consisting of his bow and arrows and a knife; his arrows, some of which (given me by Mr Bayard) are now in my possession, are stained to the feathers with blood. On receiving the order to unfasten his pony, he asserted that he could not undo a knot, and he asked the man at the trading post to lend him a knife to cut the cord. On receiving the knife he cut the cord, and jumped upon his pony, endeavouring at the same time to conceal the knife in his bosom. On this the owner of the weapon called out that the Indian had got the knife, and desired him to return it, when the prisoner tossed it in at the door, and, wheeling round instantly set off at full speed in a desperate attempt at escape.

So suddenly was this done that the men of the escort, who were not yet mounted, were taken by surprise; not so, however, was Mr Bayard. He was mounted on a favourite buffalo hunting horse, afterwards brought to my knowledge, and at once gave the rein to his smooth-going and willing steed, who in a very few strides outpaced the Indian pony, and, running round him, Mr Bayard, revolver in hand, ordered the dark rider to halt and return to custody on pain of instant death. To this order Pawnee paid no attention, but scowled defiance, urging his fast pony to its utmost speed, when Mr Bayard, passing him again and pulling across him in a humane attempt to intimidate (such attempts never answer), fired one barrel of his revolver over his head. The consequence of this was that the Indian, thinking that he had missed him, screamed a whoop of defiant derision, and went faster than ever. To permit a prisoner to escape from military custody would be disgraceful in an officer, and is, both in England and America, contrary to the articles of war. Mr Bayard, therefore, with the second barrel of his revolver shot the Indian through the heart, but as he did not on the instant fall, he fired a third barrel, the bullet of which, as the prisoner was then in the act of dropping forward, went over his head, when the redskin fell lifeless on the ground. On searching the body, it was found that to the last the man intended resistance, for though bidden to deliver up all his arms, concealed in his vest was a beautifully painted arrow, of the brightest colours, and stained at the point, perhaps with a deadly poison.

Some days after this occurrence, and about six miles beyond and west of Pawnee Fork, as the Government mail was proceeding on its road in the care of three men, it was surrounded by Indians, who bade the driver halt, and demanded food. There was something so threatening in the manner of the redskins, that while complying with their request the conductor, from the looks of the savages, saw that he was a dead man, and in a very abortive and hopeless attempt, which of course only hastened his end, ordered the driver " to head his mules for Pawnee Fork," in a desire to turn back from his mission and to escape. Of course the very moment that the mail reversed its position the Indians fired and killed two out of the three officials on the spot, wounding the third severely, who ran with his mule into the grass, where he fell down, but very luckily for him his mule continued to run away, so that the Indians, fearing to lose the beast, gave chase, and never more thought of the rider. The man then crawled away, and subsequently reached the camp of the military, where he detailed the fact as it had taken place. It was from this Indian outbreak that I met the frontier men retreating.

This narrative will sufficiently prove that I had very good reasons for not going with so small a party and so many animals into the vicinity of Pawnee Fork or Walnut Creek, which on starting from Kansas city had been the intended scene for hunting; and that my determination to seek the society and protection of the officers at Fort Riley, to all of whom I held letters of introduction, was infinitely the wisest course to pursue.

My journey, therefore, continued towards Fort Riley, and with the fever then prostrating me I longed for the shelter of a roof and medical assistance, or at least a place to rest in, where I might find a remedy and recruit my strength for the treat which, in spite of Indians, I intended yet to enjoy upon the plains.

CHAPTER XIV.

NEXT day we halted for an early dinner, and then pressed on towards Fort Riley with all possible speed, for the
fever was on me, and I longed for rest in a more comfortable bed. We continued, however, myself and men,
to pick up a few grouse. George Bromfield killed two
or three grouse very well, as also did Tom, who was a
very good shot; but they were getting very rare. During the heat of the day I confined myself to my ambulance, and had considerable reason to find fault with
and to blame Mr Canterall, for though we had many most

15

difficult creeks to cross, he was never by any chance in his place to direct proceedings. In consequence of this, weak and ill as I was, I had to get out of my ambulance, and attend to the dragging of the wheels myself; for unless I personally directed all precautions as to safety, not a man would either trouble himself to hold on to the wheels, stir hand or foot, nor attend to the due interests of the waggons. On my giving Mr Canterall a very good set-down in regard to his neglect of duty, he generally said he had gone on to look for cabins where we might procure corn, but really his object in going ahead was to inquire the way to Fort Riley, as to which route he had not the slightest conception; the fact of his ignorance, I subsequently found, was very well known to my men.

The track we then pursued was very remarkable and very dusty, the dust so excessively fine and black that we all were the colour of chimney-sweeps, particularly my men, their vicinity always being at the waggons. Our road lay through the bed of what had evidently once been a considerable river. Limestone rocks, bared on the margin of former floods, were on a level with each other on either side, while the grass grew luxuriantly on the dark alluvial soil beneath. My men said they saw traces of turkey and deer, but I saw none, and felt too ill to take much trouble about a species of game that I was well aware were very scarce.

Occasionally our course left the bed of the ancient river and crossed the plains, when, on a most open and bleak stretch of the prairies, just as it was getting dusk, from the seat of my ambulance I saw something stationary on the sky-line of a gently rising ground, and ere I could make out what animal it was, my men cried out " ante-

lope!" On this I would allow no halting, in the hope
that the antelope would continue to gaze at us, for, like
deer, they are prone to great curiosity; but, alas! a
thought on the direction of the cold keen wind, which
succeeded the hot sun, made me aware that, however
curious he might be as to appearances, his nose would
warn him of the presence of man, and therefore he would
take very good care to keep out of rifle distance. As I
expected, after gazing very hard at us, he walked gently
over the brow of the hill, and was lost to view. "*Now*
halt the waggons!" I cried; "let loose Bar, and bring
me the chesnut horse, Taymouth!" and this being done,
I set off with my rifle for the brow of the hill, beneath
which the antelope had disappeared. I knew that in my
approach he must still have the wind of me, for the ground
permitted no alteration in my line of advance, therefore
my hope was in a long flying shot, and at all events a
course for Bar. Arrived at the sky-line on which the
antelope had been standing, alas! I viewed him again on
the next sky-line, and on reaching that, then I saw his
white-looking haunches going up and down in flight at a
distance over the plains. I did all I could to make Bar
catch view, but in vain; so, it being nearly dark, and the
antelope quite out of sight, I rejoined my waggons. We
halted in the midst of the open plains, and encamped by moonlight, and a more lonely desert never greeted
the eyes of any traveller, or fostered an anticipation of
sport. On this day we killed five grouse, and I sought my
blankets, very ill indeed.

On the following morning, Tuesday, we were off long
before the sun was up, and reached a place at half-past
eight, called Clark's Creek, to breakfast. While this was
being prepared, as usual, with Brutus at my heels, and

my double shot-gun (one barrel loaded with small shot and the other with a cartridge for turkey or deer), I strolled along the adjacent cover; but so great was my lassitude that I felt scarce able to carry my gun. There was not a sign of any sort of game, so I returned to camp, and found that my men had picked some wild grapes. They were black, and not larger than peas, and acid, with very large stones. Mr Canterall, George Bromfield, and the men were better in health; but my sensations of fever and lassitude seemed to increase, though I never lost my appetite nor felt a chill, and found that the potted apples and cherries obtained in New York at Mr Duncan's, and that were in cans, refreshed me much, as did the melons and water-melons, wherever they could be procured. Brandy I left off, and confined myself to sherry-and-water and tea. Mr Canterall, who really knew nothing of the matter, insisted that the mules would never be able to reach the Fort on that day; but the only answer he received from me was " that they should try." We then recommenced our march.

In crossing one of the creeks intersecting the bed of the old river we came to a tremendously steep pitch, which, as I led in my ambulance in order to regulate the pace and avoid the black dust as much as was in my power, I had an opportunity of inspecting. In vain I looked for Mr Canterall—as usual, he was nowhere to be found; so I left my ambulance to see the drags put on, and that the men, when necessary, were at hand to hold on behind. My attention being chiefly addressed to the ambulance and to the pranks of my fiery young mare-mule, as well as to the more ponderous baggage-waggon, having seen them safely over the difficulty, I thought the lighter dog-waggon, and a fine, strong, steady mule in it, might be trusted to my men. My horror may well be imagined when, after some noise, down in the bottom of the creek, of the nature of which I was not aware, I saw emerge from it, and up its very steep side, the fine mule in a paroxysm of panic, running away, while George, who had very improperly been permitted to drive over such a dangerous place, having lost his seat on the driving-box, was still on his legs, and running to keep so, absolutely between the mule's hocks and the waggon, and still tugging, but of course without any power, at the reins. The mule, on reaching the summit of the bank, bolted out of the track through the underwood, which, luckily, impeded him, and made for a great hole, down which, had he gone ten yards further, waggon, dogs, George and all, would have been dashed to pieces. The mule saw that he must be killed if he went on, and, turning from the danger, bolted in another direction, still through thicker brushwood, when my men, aided by other impediments, succeeded in stopping him. During the whole of this struggle George never lost his head nor his legs— the one would have been as certain death as the other— but when the mule was brought up, there stood George between his hocks, or between his hocks and the cart, with reins and whip still in his hands. The cause of this difficulty was, that the place was so steep, and, I believe, from some portion of the harness or cart giving way, the driving-seat came on the mule's hocks, and shot George out of his commanding position on to the ground. None of the men, while I was occupied with the other waggons, had held on behind, and thus we very narrowly escaped an accident by which I might have lost my dogs, as well as my servant, waggon, and mule. The damage done, as it happened, was not much ; it was soon repaired, and the fine mule recovered his equanimity as quickly; but when Mr Canterall made his appearance, of course he caught it for his neglect of duty.

This fellow afterwards told me that on that day he had seen a settler who had been with a hunting party, all of whom had returned on account of the Indians, and that the buffalo were but two days' distance from us. It was very evident to me that this guide, as he was called, would have wished to have prevented my reaching Fort Riley, to which, for some reasons best known to himself *at the time*, he had a very evident dislike.

Continuing on our way, and having a very bad hill to descend, at last my men pointed to some white-looking buildings on the distant hills, and told me that that was "the Fort." About an hour before dark we descended into the narrow valley of the Kansas river, the fort being on the opposite side, to arrive at which we must cross in a very indifferent ferry-boat: so having attained the banks of the river, it being nearly dark, I ordered Taymouth to be got ready for me to ride, and some things that I should need to be put into the dog-waggon, which was to follow me, and the rest to encamp where they were, and await further orders on the following morning.

Having possessed myself of my letters of introduction, covered with black dust, and feeling the heat of the fever increased by a beard of ten days' growth (having tried that hirsute appendage, I cannot conceive any man in his senses wearing one unless in a very cold climate), I crossed the ferry, and my high-couraged chestnut steed sprang from the boat to land with a joyous toss of his head, and as if anticipating a stable he stepped gaily up the ascending ground. In my progress I met several soldiers who, I thought, looked very like our own men at home, many of whom stared hard at me in the dusk of the evening, but none of them questioned my business; when at last (the distance was longer than I thought for) I came to the top of the hill, I saw "the Fort," as it was called, but which had nothing of a fort about it, not so much as our barracks have in England, for it was not protected nor shut in even by a wall. At the two ends of an elongated square or ample parade-ground were buildings, the quarters of the officers, and on the other two sides commodious barracks for the men; a wide interval being left at the four corners, by which the parade could be attained, and which remained free to any ingress, save as the sentinels might interfere. On trotting up to the officers' quarters on the left hand, as I made my entry, I saw several, as well as one or two of their ladies, in the verandah which fronted their houses. To one of these officers I addressed myself, asking to be directed to the quarters of the commanding officer. He referred me to the buildings at the other end of the parade, adding, "Any news?" He evidently expected, perhaps from my jaded and dusty look, that I was the hurried bearer of further disasters from the redskins. "None," I replied; "but I bring a private letter of introduction."

On trotting across the parade to the quarters indicated, there again I saw a couple of officers and a lady, when, pulling off my hat, I asked for the officer in command. Major Wassells then came forward and said he was in command, and to him I delivered my letters, which, being glanced at, he said "the commanding officer to whom the letter I bore was personally addressed was absent." On this I replied, "I take it on myself to say it is free to your inspection." Major Wassells then broke the seal and read the contents, when stepping forward, in the kindest way he said, " Mr Berkeley, I shall be most happy to offer you all attention in my power." Calling to the orderly, he desired my horse to be taken to the stable, introduced me to his lady and his household, and, weak as I felt, travel-stained and tired, there was something so off-hand, kind, and gentlemanlike in his reception of a stranger, that I had to put a rough glove to my eyes for fear something should trace a furrow on the alluvial dust of the prairies. Nothing could be more kind than my reception that night, nor more refreshing than the bath on going to bed. The medical officer came to me and most kindly prof-fered advice, which I gladly accepted; when, after ten days' travel over the plains in a waggon, I consigned myself to a comfortable bed, and, in spite of fever, slept like a "top."

The doctor came to see me next morning, and said that I had much less fever than he expected; rest and a little attention, he assured me, were all that I required; but to my suggestion of being fit for a start on Thursday or Fri-day he expressed some doubt. Having had another bath and the comfort of shaving off my beard, I descended a little late for the early breakfast of Major and Mrs Wassells, but immensely refreshed; and really had my kind friends known me all their lives I could not have been received with more graceful hospitality nor attention. What if the room assigned to me was " in a rough state," as they called it, the bed was delightful after my ambulance cush-ions, and its pillow bliss; while that frank, open-hearted urbanity, so apt to the soldier and gentleman, and with which I was at once greeted, made my room a palace, and restored my health. The uniform of the army of the United States, that which I saw, was dark with a little braiding about it, and the hat of the officers, with a black drooping plume, gave rather a Spanish appearance to the head. The word " company " is applied to infantry and cavalry alike, a portion of either service being quartered in the barracks at Fort Riley while I was there.

Having been ten days on the plains in a waggon, it is scarce possible to describe the relief it was to pass a quiet day, free from the heat of the sun and dust, in the society of gentlemen and ladies, whose civility and kindness to me I shall never forget. I attended guard-mounting, and with Major Wassells inspected the barrack-rooms and arms. The same drill as in England was evinced when " 'tention" was called on our entering the barrack-rooms, and at first for the life of me I could not make out why the privates of the companies reminded me of those I had been used to, more or less, in England. This was very soon explained, however, for they were all Irish, with the exception, I believe, of a few Germans; and there was nothing in that part of an army before me that was different from ours, save that the smart and superior class of men who officered it were all of them Americans. In mounting guard they " troop" much as we do, but the officers in mounting and relieving guard exchange more words, while their salute with the hand is made by flinging out the arm from the chest instead of up to the forehead. The men went through their exercise, both sword and musket drill, very well, and, considering that the enlistment is but for five years, I saw some well-drilled, properly set-up soldiers, and some very smart non-commissioned officers. The barrack-rooms and beds were neat and in good order, but, though the officers took care that the arms were well kept and clean, a great many of them were not fit to be relied on in service.

The lassitude which beset me, and my desire to recruit my strength sufficiently, made me retire to bed early,

which I was enabled to do through the most kind but
unceremonious attention by which I was surrounded,
when about ten o'clock on the second night of my arrival,
just as I had fallen into the first deep sleep, I began to
dream of home and all the loved scenes that neither time
nor distance could make me forget, and once more even
the national airs of my country were in my soul as if they
had been actually played. The airs grew on me still
more—my eyes opened, and in a dreamy trance my ears
had been serving me faithfully, for under my window,
plaintively or gently suppressed, the regimental band was
serenading, and pretty English melodies were really
floating on the night breeze and gracefully speaking of my
country. The real soldier and gentleman of the United
States—who knew how to wear a sword and use it, and who
was prepared to work for and to win the distinctions of
rank, so utterly different from the self-constituted generals
and majors I had so abundantly met with in my travels—
many of them hearty good fellows but no soldiers—was in
America precisely what he is in England; and between
the officers of the one country and the other, in mind,
bearing, and bravery, the difference was nothing. While
on this visit to the barracks, the chaplain to the Fort came
to see me, and I had the satisfaction of shaking by the
hand a countryman of mine, and a Bristol man, and it
gives me very great pain that at this moment I should
have forgotten his name.

My wish to continue my journey on the following
Friday having been repeatedly expressed, I was told that
if I would accept their companionship, two of the officers,
Lieutenant Bayard, of whom I have before spoken, and
Major Martin, would accompany me; but as "Friday
was an unlucky day," they wished me to remain at the

Fort a little longer, and that at the end of the week they
would be prepared to start, with a leave of absence for
eight days. I believe they said that Friday was an unlucky day more because, out of kindness and friendship,
they wished me to rest a little longer, rather than that they
really had any feeling of superstition, and the result was
that I agreed to accept of their hospitality till Saturday,
the delay suiting my brown mare Sylph as well as myself,
as she, too, had been unwell. I must not forget to mention that on its being known to him that I was out on the
plains, in case I should need refreshment and rest, Mr
Wilson, the author of a clever work on the "Conquest of
Mexico," in the kindest and most hospitable way had
rooms in his house near the barracks prepared for me,
to be put at my command for any time I pleased. This
fact I was not aware of when I delivered my letter of
introduction to Major Wassells, but having been received
at once so gracefully and kindly by him and his lady, of
course I could not desert my quarters.

My time at the barracks passed very pleasantly, and
the more I saw of the society there the greater reason I
had to be pleased; but though I kept a journal, and now
write for publication, I never chronicle the transactions
of private life. Without a wish any further to arouse the
lion press, in which, according to Dickens's "Colonel
Diver of the Rowdy Journal of New York," "the enlightened means and bubbling passions of his country find a
vent," I must here take leave to say, that "the Jefferson
Bricks" are already trying, "through the mighty mind
of the popular instructors," to "rile" the people against
me, although in two public lectures—one of them, on a
rainy night, I am told, the most numerously attended that
the great room at St Louis had ever seen—the same

people had applaudingly backed the very sentiments that
these "Divers," "Chokes," and "Bricks" are now so
sedulously condemning! How the "Jefferson Bricks" of
Boston (*cum multis aliis* in the United States) in their un-
fair attacks on me, and by their vulgar personality and
abuse, bring to mind Lord Byron's cuckold in "Don
Juan," and essay to "prove themselves the things they
most abhor." I can't say much for the discretion of
these military-titled editors, while, at the same time, if
America really desired to repudiate the descriptions of
Dickens, in his "Martin Chuzzlewit" and other works, she
might well and truly say, "Stop these New York stabbers,
and the family spies, private listeners, peepers, plunderers,
keyhole reporters, rowdy journals, at Boston and in other
cities in the States, and *only* 'save me' from my friends
of the press, and 'I'll protect myself against my enemies!'" For the life of me, I can't help laughing at the
riled-up or "enlightened means" by which, according to
"Boz," in "Chuzzlewit," p. 194, "the bubbling passions
of America find a vent;" and, again, quoting from
"Diver," soothly will I say, and with the utmost com-
placency, as well as forgiveness of all harsh language,
"arter" these outbreaks of fury "let's have a glass of
wine."

To return to my narrative;—the two officers who were
to accompany me, with their horses, men, mules, and waggons, were Major Martin, a gallant soldier, who had lost
an arm in his country's service, and who had been the
recipient of a complimentarily presented sword, and my
young friend Lieutenant Bayard; the former to look on
at our sport, as the loss of an arm incapacitated him from
otherwise sharing in it, and the other as one of the best
and most successful buffalo or bison hunters in the United

States army. Not only was I well pleased with my two friends in themselves, but I was also charmed that I should have an experienced bison hunter to start with side by side, and witnesses, too, of what took place, for some of the "rowdy journals of the American press" had already begun to try to sneer at "the English lord or baronet" —the ignoramuses always gave me the one title or the other, and often both—" who had boastfully expressed his resolution to step across the Atlantic and walk into buffalo in shorter time than it had ever been accomplished by man ; " and I well knew that, with a departure from truth much abroad in these realms, the mere vicinity of the Rocky Mountains is fatal to an approach to veracity in Englishman or American, all sorts of falsehood would be invented by scurrilous detractors, and that I should, by an immense portion of the press, and in the tyrannizing spirit of the realms, be denied fair play.

In this, my first visit to the Fort, I saw in the possession of a soldier several of those animals called "prairie dogs," one of which I subsequently procured, and which has since been placed in the Regent's-park Gardens.

On Saturday morning, the 8th of October, my dog-waggon came up to barracks to fetch my goods to the camp, which, according to my orders, had been brought across the river and pitched nearer to the Fort, whence the officers as well as myself could amuse ourselves with paying Druid a visit, and then we proceeded with the Fort Riley waggons to join forces and direct our march for the nearest spot in which we hoped to reach the bison. A goodly company then we formed, for the officers had, if I remember rightly, ten or eleven well-armed soldiers with them, carbines and revolvers all complete, and two waggons containing provisions, their tents, and an officer's

tent and bed also for me. Major Martin brought with
him his little boy, William, with whom it was impossible
not to be pleased, and a scrap of a song I sang to him, of
" Hoot awa', hoot awa', wandering Willie," became the
tune that everybody hummed while in his hearing. When
his own pony was not fit to go mine was at his service,
and when he did not accompany his father to the plains
he caught small chub and other fish in the creek near
which we encamped.

On setting out upon the march, I found that I was not yet
equal to much exertion; however, we killed a few grouse,
and then after a little time I was obliged to seek my ambulance and lie down, for though actual fever was not
still on me, yet the enervating attack had left me almost
overcome by an inexplicable lassitude that it was very
difficult to shake off. We camped that night at a place
called " Chapman's Creek," computed at about twenty-
five miles from the Fort, when, seeing that there was a
swampy portion of it that looked likely for wildfowl, I
took my gun and Brutus, and, on two bitterns rising, I
killed the first, but my gun missed fire at the second.
There was an island where this took place, and I
observed that there were large paths across the sort of
green duckweed that lay on the surface of the stagnant
water, which must have been made by creatures larger
than the musk-rat, and therefore perhaps by beavers or
otters. Had I had the strength to have done it, I should
have explored that island, but, for the sake of the grand
affair expected so shortly to come off, I was glad to get
my dinner and to go to bed in the comfortable tent the
officers had pitched for me. On coming into camp, alas!
the ill news reached me that one of Mr Bayard's best

buffalo horses—he had still a beautiful one left—had fallen so lame as to be, for the present, useless.

The tents, those of the officers and myself, were always pitched close together, and we made mine our dining-room, the ambulance and dog-waggon drawn up close to me, that I might be near my favourite dogs, have my arms and ammunition close to me, and George Bromfield as well to wait at dinner, of which duty he acquitted himself admirably. Now, for the first time, I had very clear evidence of how distasteful my league with the officers at Fort Riley was to Mr Canterall. That fellow pitched his waggon and tent as far from that of the soldiers as possible, and unless he had the most direct orders from me, which he knew he dared not disobey, he never brought anything from my kitchen to the general mess, the two ends of the encampment beautifully illustrating the two expressions, discipline and disorder. Our dinner table, however, on the first night, afforded game and a bit of beef the officers brought with them, as well as potted meats, potted salmon (which latter was excellent), and sardines, sherry, brandy, whiskey, coffee, and tea; so we were very jovial and comfortable. After dinner Mr Canterall came into our tent for orders—at least he made that the reason for intrusion, and usually took a seat, with the accustomed familiarity of his country, on the corner, at the foot of my bed. When there, it was amusing, though disgusting, to see him spit in spite of himself, for he was perpetually oblivious of my orders in that particular, though wishing to obey them; but habit was all-powerful, and the rascal was therefore never at his ease. He came to my tent in reality to get what he was very fond of, even, as I subsequently found, to its theft,

—brandy, but, it neither being good for the state of his health nor according to my pleasure that he should have it, he obtained only that which he pretended to desire, the orders for next day, and with those he was coldly dismissed.

If I found the barracks at Fort Riley a sort of heaven upon earth in regard to the habits of much of the society into which I had latterly been thrown, and had discovered that, with some exceptions in other places, the army alone held the courtesies of English life, and the abstinence from the almost universal vice of promiscuous spitting, my pleasure was again agreeably enhanced by the discovery that my two companions, Major Martin and Mr Bayard, never used tobacco in any one way, and that therefore I was not bored with even the fumes of a cigar. Nothing, then, could have suited me better than the society into which I had been thrown, when, after a good night's rest, by eight o'clock on the following morning we had done breakfast, and, much refreshed after the cool of the morning was over, I again rested in my ambulance, while my friends rode on a little ahead of the line of march.

Mr Bayard had brought with him a greyhound, with whom and Bar we hoped to have a course at hare or deer. We were now fast approaching that part of the plains where report had brought us word the first herds of bison were to be met with, or, in other words, that we should find them "two days' march beyond Fort Riley;" and in consequence of this, every eye was occasionally addressed to the undulating hills and to the various sky-lines, and many a dark stone or spot in the landscape anxiously scrutinised. For myself I felt so wild a longing to see my first bison that I could think of nothing else. Taymouth was ordered to be led close in the rear of my ambulance, and my Prince's carbine with its cartridges lay ready to my hand; whenever my curiosity was attracted to any suspicious-looking objects among the russet undulations or hills on the prairies, for a moment all lassitude seemed to leave me, and I felt in my accustomed strength and elasticity. For some time the solitary monotony of the wild landscape lay in its enormous and unbroken extent around our cavalcade, and there was not a semblance of other life than that which we brought with us to be seen. All at once Major Martin and Mr Bayard, who were riding at a considerable distance ahead, halted abruptly, and while the Major and Master Willie stared intently in one direction, Mr Bayard looked towards my ambulance and held high aloft his hat, conveying to me a spirit-stirring information, the purport of which I felt there was no misunderstanding—he had sighted the great bison on his native wilds !

CHAPTER XV.

IT was about noon, when Mr Bayard made that im-
possible-to-be-misunderstood signal with his hat, as
recounted in the foregoing chapter; so, while my horse
was being brought up, and my spurs buckled on, I knew
that (for the first time in my life) I was about to have a
run at the giant game! Oblivious, then, to every unpleasant sensation that the fever had left, I mounted
Taymouth, Prince's breech-loading carbine in hand, and
joined Mr Bayard and Major Martin, and very soon set
my longing eyes on about thirty huge old bull bisons
grazing quietly at the distance of a mile. Beyond them,
again, there was a much more numerous herd. Prepared
as I had been to see a large animal, these bulls loomed
infinitely more magnificent than my fondest imagination

could have depicted, and, instead of being lost or
lessened in the infinity of space around them, they stood
forth out of it in such black, bold relief, their long
manes and beards flowing in the wind, that in size they
seemed to resemble elephants more than bisons, and
were, indeed, to me a most novel as well as a splendid
picture of the largest and wildest-looking game!

Bayard and myself (I confess to have been in a
charmed delight) then set off towards them gently and
without noise, availing ourselves of any inequality in the
ground there might be to cover our approach, and in
order to give the bisons as little the start of us as possible ; but when we came to within about half a mile of
them, off they set in that peculiar up-and-down canter
in which they invariably commence their retreat. This
canter of theirs gains its appearance of height from the
hump on the shoulders and the tossing up of the long
mane, more than from any high action in the legs. The
instant we set off at a gallop in our run to the game,
Taymouth was all on fire to keep ahead, and when he
saw the retreating mass of beasts flying from him, ignorant of what they were, it increased his anxious desire
to overtake them. Having heard that horses were terrified at even the smell of as well as the sight of bisons,
I drew Taymouth into the wake of the retreating animals, in order to encourage his approach, and to let him
know they were in retreat. He soon overtook them ;
but when he came up to within about fifty yards of the
rearmost bull, while he slackened his pace a little, he
pricked his ears and made such a stare that I knew, as
well as felt, he was very much scared, and inclined to go
to the right-about. A slight touch of the spur, however,
and that clasp of the knees which horses so well under-

stand, put all direct refusal out of his head, and we came up at three parts' speed alongside the bulls, though he swerved from them infinitely further than I desired. But for the rein and heel, he would have gone clear away; with these adjuncts, I managed to keep him, though fighting against me, to within about twenty-five yards of the bulls. His so much over-pacing the bisons enabled him, when restrained, to change his legs and fling his head and shoulders in such a way from side to side, that to aim and fire with the carbine was for some time impossible. Oh! what an exciting wild sight it was, thus close up with them, to see these thirty rusty black monsters, flying two or three abreast, or else close in each other's wake—the last old bull (generally the king of the herd) leering out from side to side beneath either horn, as much as to say to the pursuer, "I *don't* like you, and I *am* retreating; but just you get into *my* way, that's all, and then see what I'll do." Bayard, I am sure, did not run for a shot at first, himself, but rather waited to observe me, for not until a large bull that I had pressed left the herd was I aware that he was close to me, and then too I saw that Mr Canterall also was in company. As Bayard seemed to be holding back for me, I called out to both of them to go at the bull, when Bayard, on his steady nice horse, ran alongside, and, with his heavy revolver, slightly struck the bull, but not in a spot to stop him. The bull then became mischievous, and prone to charge anything that came in his way—of this he made both Bayard and myself well aware—and as bisons often do, when stricken or in a fighting humour, he took no more notice of the direction of his herd, but went away sulkily by himself. I shot at him without effect, and then Mr Canterall, as I found

Printed by.]   [Spottiswoode & Co.

THE FIRST RUN AT THE BISONS.

afterwards was his usual custom, fired two or more shots
at the bison with an army carbine, and missed him
clean, my horse Kansas, which Mr. Canterall rode, being
beautifully steady.

We now came to a creek that intersected the plains,
down the steep bank of which, without the least pause in
his long gallop, the buffalo went in the oddest and most
reckless way I ever saw, getting a complete summersault
into the water at the bottom, at which I was immensely
amused. Bayard and myself then halted on the brink of
the creek, and waited for a steadier shot at the bull as he
climbed the other side. Bayard fired with his revolver,
at a long distance for that weapon, and I got my second
shot, and saw that it took effect in a slanting direction
on the back of the bison. We then rode over the creek,
and my third shot, at some distance, broke the shoulder of the huge beast, proving the strength of the
shooting of Prince's carbine, and brought him at once to
bay. We then drew up at a respectful distance, as the
victory was sure; the monster, lame as he was, being
ever ready to charge, when, drawing a little closer,
Taymouth being somewhat quieted by the length of the
chase as well as his fractious exertions, I opened my left
side for the facility of a shot, and hit the bison close
behind, and a little above, the elbow, when he swayed
from side to side for a moment, and then fell dead.
Making much of my horse, I rode him up as near as he
would go and dismounted, when, giving my hunting-
knife to Mr Canterall, I bade him cut off the peculiarly
immense beard of this bull, which, with his tail, now
ornaments my rooms at Beacon Lodge—the tails of the
bisons, handsomely set in an acorn the size of a small
pine, by Harvey, of Lambeth House, Westminster-bridge

road, append from and make a sporting finish to the bells of drawing-room, dining-room, and study. Mr Bayard then, in the most scientifically sylvan way, took out the tongue, while Mr Canterall availed himself of some of the meat; when, as we were too far removed to get either his skin, the best of the meat, or his marrow-bones to camp, very reluctantly I left such a waste of good things to be eaten by wolves, or to enrage the aboriginal redskin of the soil at the wasteful destruction, by the white man, of the animal from whose herds were derived the chief subsistence of Indian tribes.

A more splendid beast of chase than the one in question could not be, nor can there well be an animal of greater muscular power and bone than is the bison on the western prairies. Added to an enormous depth of carcase, the ribs are immensely expansive, while the quarters of the animal are indeed "rounds of beef," as circular as they can well be. His hocks are peculiarly strong, while between the hock, knee, and fetlock, the leg is short, and the bone very large. The head, neck, and hump are so very heavy that the animal gives a sort of lift with his head to assist the action of his fore legs when he starts into a canter. By way of proving the immense width of chest and roundness of figure that the bison has, I cannot do better than tell the reader that in all the bulls we killed during our hunt, when lying dead on their side, the uppermost fore leg stood straight out, in a direct line from the body, the foot not even inclining to the ground. During this run at my first bison, in crossing a small creek of water, Mr Canterall said he rode over an otter in the rushes. He might have done so, but, as I crossed within ten yards of the same place at the same time, and saw nothing like an animal of any sort, I do not believe it. It was impossible for this fellow to stumble on the truth, even by accident.

After the excitement of this splendid run was over I began to find a return of great lassitude and stiffness all over my limbs ; but, as we returned towards our waggons, we saw another herd of buffaloes at a considerable distance from us, and on ground of which we could take no advantage in our approach ; the consequence of this was, that when we were scarce within a mile of them, that peculiar toss of their heads and apparently high action denoted that they had seen us, and were off. Between us and the herd there intervened a little rill of water, in the midst of high grass and rushes, the latter some thirty feet broad ; the little creek itself, scarce four feet wide, but worn into the soil perhaps five feet deep, and lapped over and completely disguised into a sort of pitfall by the grass and reeds that grew so many yards on either side of it, was completely hidden. This obstacle intervened just at the start ; Bayard was leading, and his steady horse walked through the stuff, put his head down to peer into the water, and got over ; Taymouth, however, now that he had been made aware that a race was the order of the day, on seeing the grass, and guessing at the obstacle, for he could not see the water, and knowing that Bayard was on ahead, tried to rush blindly after the other horse, which I prevented, as to let him thus rush into it was to come to perfectly certain grief. On being checked, he turned his tail to it and commenced restiveness and a succession of jibbings and rearings, to which I had to drop my hand to prevent the chance of his going backwards. At this time I felt so giddy and weak that I could hardly retain the saddle, and was at last obliged to dismount. After a

little time Mr Canterall and one of Bayard's men came up,
when I led my horse through the obstacle, and then, with
their assistance, remounted, but I was much too exhausted to take up the running, even if there had been time for
me to do so.

Away in the hills I then heard Bayard's heavy revolver,
when, on proceeding steadily towards the spot whence
the report proceeded, a wolf crossed me, but on pointing
my carbine, Taymouth, now knowing what that meant,
plunged so that he prevented my fire, and constrained me
seriously to vow that the day should come when I would
try to cut out work enough to make him glad to stand still.
I then found Bayard standing over the body of a fine
four-year-old bull bison, up to which his steady, splendid,
and smooth-going buffalo horse had taken him as straight
as a line, and, if I recollect right, he disabled the bull at
the first shot, but it charged him very viciously ere he
received his death-wound.   Before he killed the bull he
made a dash at a heifer, but she shut him out in the ruck.
This fine young animal being pretty fat, Bayard and Mr
Canterall set about taking as much of his meat as they
could carry, and I took his horns, as they were very
smooth and evenly grown.

Pretty well laden with spoil we were on our way to
camp, when we saw several bulls at feed, one of whom
was so situated that he might be approached on foot. To
this bull I approached to about a hundred yards, and shot
at and hit him, as my ear told me, but he went off, as
bisons will go off, if not hit in some spot that is almost
fatal, and I was too weak and tired to attempt the run. On
reaching camp we found that some bisons had crossed so
near the line of march that the men had had a volley at one,
and had succeeded at first in only wounding him and then

in killing him, so that my servant George Bromfield had
helped to despatch a bison for the first time in his life.
We found the camp pitched on a creek of running water,
on which the waggons were ordered to march when we
commenced hunting, but they had selected a spot not
very well provided with fuel. Of beef we had now plenty,
for on this the first day of sighting buffalo we had killed
three, besides a few prairie grouse picked up in the morning. From first to last I had also viewed two wolves. We
needed a good supply of fresh meat, for in all we were,
if I recollect rightly, four or five and twenty men.

On reaching my tent, I divested myself of my belt and
ammunition, and lay down for half an hour on my bed,
drinking at the same time a large glass of very fair
sherry, and before dinner was ready I felt greatly refreshed and as hungry as a man could be who had no more of
a fever left than weakness and the appetite which such
attacks when surmounted very often occasion. I was still
in my tent when little Willie came to ask me to come and
look at Bayard's horse, who was very lame, and I hastened out accordingly.   To my deep regret I saw that
nice steed with a knee swollen to twice its size, the effects,
as I at once declared, of a severe kick. "Oh, no," the
men said, "it could not be a blow, it must be a strain ; "
of course they said so, because their orders were, never to
let the animals reach each other from their respective picket-
pins. We, Bayard and myself, however, decided that it
was a blow from the heels of a mule—they are always
handy in that way—so orders were given for an immediate
hot fomentation with vinegar. Mr Bayard also bled his
horse from the mouth, breathing a vein elsewhere not being
understood. We then dined in my tent, and did ample
justice to bison soup, broiled hump rib, grouse, and other

good things, washed down with sherry, brandy, or whiskey, which we liked best, and talked of the events of the day, till the time for rest arrived. Brutus slept in my tent that night, an indulgence of which he was very fond.

Having arranged my bed, and in a chair at the head of it put my revolver, breech-loading carbine, and Pape's breech-loading double rifle, all loaded and ready for use, feeling heated, I resolved to go my rounds to cool myself, as well as to ascertain that my sentinel over my portion of the camp was doing his duty. The first thing I came to was a loose mule, which had pulled up her pin, and the next two mules put so near together that they had got their larriets entangled. I went then to the watch-fire, where one of my men named Tom was supposed to be on the look-out, and roused that gentleman from a listlessly sedentary attitude, and asked him "What use he was of, if he could not attend to the animals during his watch?" On this he replied most insolently, so I at once told him if he gave me any more of his impudence I would that instant turn him out of camp, pay him up to the day of his discharge, and leave him to find his way back to Kansas city the best way he could. He said something about a constable when we got back if I did; but I told him that out on the plains there were no constables but myself, when, on finding that I would put up with no sort of American independence, he became more civil, and said he did not wish to displease me, and the mules were then forthwith better attended to. After this little episode in the history of a prairie camp, I went to bed, and, in spite of the yellings of wolves, slept till daylight the next morning, although the night was disagreeable, cold, and windy, and my tent had not been well pitched; it was too slack, and the wind not only flapped it about, but, though I had taken the precaution to tread down the long grass within into a sort of carpet round its walls, sundry draughts came up and found their way to my head in spite of my endeavouring to bury it in the pillow. It did not prevent my sleeping, though, and on the next morning, after broiled bones, hot coffee, and some nicely-baked soda rolls, also hot, I felt considerably stronger than I was the day before.

In the chase of the previous day I had seen enough to know that Bayard was a fine and resolute horseman, as well as a good shot, and that whatever I might succeed in doing there was no beating him; it would be quite as much as I could expect, on fractious and unsteady horses, to be anywhere by his side. On inspecting his horse, for the second time a terrible ill luck attended his stable; the swollen knee, though slightly better, still offered so severe an impediment to work, that it was resolved to send both his lame horses back to Fort Riley : so our party sustained the loss of a man, and Bayard the use of his two best steeds. My bay horse Kansas, which Mr Canterall had been riding, was then put at the service of Mr Bayard.

Ere we left camp on Monday, the 10th of October, a considerable flock of wild geese passed over our heads, flying high, although the wind was very strong. On this day I rode my brown mare Sylph, and suspecting, in spite of her beautiful and temperate temper, that she might be afraid of the bisons, I asked Bayard to lend me a heavy, six-barrel, army-revolver, such as the one he used, as I thought that it would be more handy on a terrified horse than a carbine. On this day we desired some men and a waggon with six mules to keep us in sight, in order to bring home such trophies as the sport afforded ; and we were not long in viewing a large herd of bisons, but at some distance; they soon saw us, when just as we started at them, and were going at full speed, Bayard riding a horse of Major Martin's, again one of those old winter watercourses intervened, grown up with long grass, but having no great rill of water in the middle. Bayard and myself came at a close together; I kept my mare well in hand, but he went into it at tip-top pace, and by the continuation of his speed was a few lengths in front of me, when I was pained to see his horse, from the blind inequality of the ground, throw a complete summersault on his rider, who was cumbered with a second heavy pistol iu his belt, and then, in rising, tread on his hand and leg. Sylph, held well in hand, made no mistake, and ere Bayard's horse had gone twenty yards I caught him, when, on seeing that Bayard, though apparently shaken, was not seriously hurt, and that Mr Canterall had come up, I handed him Mr Bayard's horse, ordered him to attend to him, and then gave Sylph her fling at full speed to overtake the herd of bisons, who had thus gained a considerable start. How pleased I was with her fast smooth action, though she was neither so fast nor so well up to my weight as Taymouth, and oh, how beautifully she gained and reached the herd, consisting entirely of large old bulls! but when she began to close with them, at first she was more frightened than Taymouth. However, I forced her to within twenty-five yards of the long string of retreating beasts, and delivered my six barrels from my revolver, as close as I could induce Sylph to get, to a sly-looking old bull, who more than once slackened his pace with a very wild leer at me, and that ominous crook of the tail as I came near him which ever portends a charge. I could not see that my shots took any effect, though I felt sure I had hit him, but the noise of a hun-

dred bulls in their gallop over the prairies prevented my
distinguishing a hit by ear; however, the bull suddenly
gave indications of distress by slackening speed and
leaving the herd, and I am sure that one or the other of
my shots had done the office. At this moment, and when
I had fired my last barrel, I was delighted to be joined
by Bayard, followed by Mr Canterall, when the latter
missed the bison with an army carbine twice, and Mr
Bayard fired at him and hit him hard.

On reaching a small creek, the bull there stood stock
still and turned to bay, though neither myself nor Bayard
could see where he was wounded; nevertheless, it was
very obvious that the game was dying. Mr Bayard shot
at him again, and I took the carbine from Mr Canterall,
and fired at him twice, Sylph standing fire rather more
steadily than Taymouth, when the great bull reeled
slightly and dropped down dead. Having taken his
tongue and tail, &c., we returned to the waggon, deposited
our spoils, and then, finding that Sylph was more man-
ageable than Taymouth, I took in hand my double
breech-loading rifle, made by Pape of Newcastle, the
cartridges ready capped with the needle, so that it
loaded without the least additional delay; and Bayard
mounting my bay horse Kansas, we set off to search for
further game. Bisons were all round us, and we soon
viewed nearly two hundred in a herd, and on them we
immediately ran. The herd separated into detachments,
and myself and Bayard ran buffalo in different directions,
while Mr Canterall followed other beasts we knew not
where. The section which I followed and came up with
was a large one, and I passed to the right of the string
at about 30 yards, Sylph at that distance giving very
little trouble, and then, when I came up with the leading

buffalo, taking very good care to watch that the rear of
the route did not deploy and get me in line, I pulled up,
and thus got, if an unsteady shot, still such a shot as enabled me to take some aim and hit one beast or the
other. My first shot told, but struck the bull, as I very
well saw, too far back, or, as we should say with a deer,
paunched him; my second shot I could see nothing of,
and then I slackened speed to reload. Alas! the exploded
cartridges neither of them would withdraw, by the purchase of the exploded needle. My hunting-knife was too
large to permit of a hold on the edge of the cartridge
itself, and I pulled up, terribly chagrined at seeing the
herd, with the wounded one falling to the rear, keep
on in their persevering gallop till they got out of my
sight. There was nothing left for it, then, but to return
to the waggon.

In going back, however, my eyes detected Bayard
standing over a two-year-old bull, which he had killed,
and Mr Canterall also joined me with an assertion that
he had killed two large bulls and a calf. On my asking
where they were, he replied, "Oh, there away, that's a
fact, guv'nor, t'other side the hills." "Where are their
tongues and tails?" I asked. "Oh," he replied, "I did
not stay to take them, I came on after you." "Very
well, then," I said, "*I believe the calf*, for at starting I
caught a glimpse of that; but as to the bulls, I don't
believe a word of it; come on to Mr Bayard." We then
went, and got the excellent meat, &c., of the two-year-
old; that done, in some amusement I asked Mr Canterall
"to take me to his two bulls." To that desire he made
evasive answers; first "they were too far," then "they
were very old, and not worth going after," and at last he
abandoned the idea of them altogether, the truth being

that he had hit nothing but the calf. The whereabouts of the calf I knew myself, so, followed by one of my men, I think by Philip Smith, one of the best of them, carrying my old favourite muzzle-loading single John Manton rifle, I went to look for it. The calf was only wounded, and had lain down in some long grass, and I rode by the spot without seeing him; not so my man, who was on foot with my rifle; he nearly trod on the calf, which, when it got up to run away, he fired at and killed, for which he got a good reproof for using my rifle without orders, as I could have ridden the calf down without the waste of ammunition.

The waggon having been laden with the spoils, we turned our heads towards camp, and very soon after beheld a fresh herd of bisons, full two hundred strong. Bayard and myself and Mr Canterall immediately charged —they were considerably to my left, and went at a wing of the herd in that direction; while I, seeing the inclination of the leading animals, and having Pape's double rifle in hand reloaded, with a penknife to disengage the cartridges after future explosion, was making my way to the head of the retreating column, in the hope of some part of the string passing me, after I had reached the front and come to a momentary standstill. Sylph was at tip-top speed flying in this direction, when, some way ahead of me, only to my right hand, I saw rise from his lair in some longer grass than usual a beautiful male antelope, who, not seeing me, fixed his full eyes in a stedfast gaze on the ruck of buffaloes and their two pursuers. A very little inclination in my line would bring me within distance of the new game, while at the same time I should lose very little of my position in regard to the bisons, and I hoped to escape the notice of the antelope till notice would be too late. Before I reached within shot, however, the beautiful beast turned his head, looked at me, and then darted off at tip-top speed in a contrary direction from the bisons. To follow him was useless, so, never having slackened speed, I again went at the buffalo.

Just before I reached them, a calf sprang out of some grass in an endeavour to overtake the herd, and it being all in the line I put Sylph to her utmost speed to get a shot at the calf. At first the calf was on my right, which was not easy for a rifle shot, so I charged right on the little game, Sylph more steady, and not caring for a creature of that size; when I was immensely amused by this buffalo in miniature, when she found I was almost on her, drawing up, crooking her tail, and charging right at the mare, narrowly missing her quarters ! The calf thus passing behind me, then came up on my left, which was just where I wanted her, both going at full speed, when, with the first barrel, I rolled her over with a bullet through the heart. By this time I had heard several shots at the herd, so thinking that there would be wounded bulls to deal with even if I delayed a few minutes, I jumped off Sylph and took the tongue and tail, without the production of which we made it a rule never to believe in the death of game. Major Martin saw me make this shot, which, as a successful attainment, was a brilliant one. I was soon on Sylph again, and to my great joy met eleven old bulls coming back upon me. On seeing me they veered off.   To these I immediately gave chase, and reached them close enough to deliver two shots, but without any visible effect, when, on finding that Sylph, who had had a very hard day, so soon after her illness at Fort Riley, was very tired and unable to continue the pace, I pulled up, and went slowly in the direction of Bayard. In meeting these scared buffaloes retreating from Bayard, I at once learnt that when we are in chase of a large herd the herd never goes as fast as it can go, and for this reason: the leading buffaloes, always the cows and calfs and barren heifers, are not at first aware of the immediate presence and pressure of danger, when, as the bulls always follow in their wake, taking the pace from them, and acting as their rearguard, we overhaul them easily in a hand-gallop. Not so with these eleven old bulls—they had been made aware of danger; and it was with a great struggle that Sylph, tired as she was, outpaced them; and as to my pony, which my man Phill rode, carrying my stalking rifle in case I should need it, he found that he could scarce keep near enough to see which way we went, and expressed his wonder at the speed of the buffaloes.

Soon afterwards I came in sight of Bayard, standing by the body of a fine old bull, excessively disgusted with Mr Canterall, who as usual, having failed to kill a bison himself, came up just as Bayard had mortally stricken his bull and brought him to a dying bay, when, instead of leaving Bayard to finish him, Mr Canterall shot into the foe already half slain, and seemed to think he had done a clever thing. As to this, he got a caution from me for the future. The etiquette among hunters is, never to interfere with an already safe animal unless told to do so. It was dark when we reached the camp, horses and men all tired; but my strength improved, though the lassitude from which I had suffered for some days still hung about me, but it did not prevent my enjoying my dinner. At night, before it had long been dark, the

17

chorus of wolves began, to thank us, perhaps, for the dinners we had left them on the plains; and ere long the
muttering of thunder and a pitchy horizon, its jet black-
ness advancing against the wind, occasionally enlivened
by forked lightning, warned the camp that a terrible
storm was coming. It came! with wind and such rain
and hail, lightning and thunder, as I had even never yet
seen nor heard. I thought my tent would have been
blown away, while the deluge of drenching rain sounded
precisely as if my locality had been situated under a portion of the Falls of Niagara. I had a light in my tent at
first, and it amused me to see the effect the storm had on
Brutus, and his awe-stricken look. He sat bolt upright
listening to the elements, and occasionally turning his
eyes anxiously on me to ask for explanation. Storms and
unrest, however, cannot last for ever, and with the retreating thunder, but still amidst the rush of the descending rain, I fell asleep, but not to a very refreshing slumber, for I was over-tired and again prone to feverish
sensations, so that sleep came to me by fits and starts.
Pour, pour, pour—spatter, spatter, spatter—the rain continued all night, and in my waking moments I had very
grave doubts if the state of the plains would permit us the
next day to go in search of what the Americans call elk
on the Far West plains, but which are really the American red, or the wapiti, deer, as proved by the shape and
nature of their antlers. The bisons, of course, drive
away all other grazing animals, precisely as with
other kinds of game in many other countries. The
hunter would not be well skilled in his craft if he expected to find the buffalo and wapiti together, more
especially as the wapiti are very fond of woods.

During my range over the plains up to this time I

THE WAPITI.

had kept an anxious watch for the prairie dog, as I was most anxious to see what sort of creature it was, but all in vain. Several men who *professed* intimacy with the Far West, to whom I had been referred as experienced hunters on the plains previous to my leaving England, had replied to my question in affirmation that " they had been acquainted with the animal," and that " oh—he was a sort of little dog ; " but further than that they could not tell me. While in America my curiosity was still increased by a statement which I received and still receive with caution, that there was a small species of owl that lived in the holes with the prairie dogs, and that rattlesnakes did so too. Now, if this tale be true, were the snakes and owls in predatory search for the young of the prairie dog, or is the " town," so called, the resort of a " happy family ? " and in case of snakes " in their little nest *dis*agreeing," has the prairie dog, or " marmot," of which class he certainly is, a power of resistance to snake-poison ? I am trying to obtain a further supply of live prairie dogs, and it will be an experiment for Mr Buckland, to let one of the rattle-snakes, — I think there is one in the Regent's Park Gardens,—if he can obtain permission, be put with the prairie-dog, to see if there be any fellowship between them. Before complete familiarity be allowed, however, between the snake and marmot, the loss of the first must be prepared for as well as the latter, *for very great hostility between these two may exist*, in spite of American belief that they are friends, and the marmot has certainly the power to kill a snake *if he possesses a similar knowledge to the ichneumon.*

CHAPTER XVI.

On the following morning, the 11th of October, the
storm that had commenced with heavy thunder and
lightning on the preceding night abated not its continuous rain, and a more dreary prospect than the prairies
afforded for travel or sport could not well be imagined.
Bayard's two lame horses being for the time of no further
use, were therefore ordered to return to Fort Riley on
the next day; and, on finding myself again oppressed
with feverish lassitude, simply from over-fatigue occasioned by exercise and the effects of illness, I resolved
to keep my tent, and to rest myself till better weather.
An excellent breakfast of very good and very hot coffee,
perfectly charming in the cold, damp state of everything
around us, flanked by broiled eggs and ham and a steak

from the loin of the young bull, with some of the "consolidated milk" brought in the sealed cans from Mr
Duncan's store in New York, to mix with our coffee, set
me considerably up again; but as the weather was bad,
and my running horses wanted rest, I determined to
remain quiet. It is well that I should here remark for
the information of future travellers, that the consolidated
milk is good, but the granulated milk is not so.

Bayard, taking Mr Canterall with him, each provided
with a picket-pin and larriet at the saddle-bow, and
rifles—Bayard having my double Pape's rifle always at
his service when I was not using it myself—they started
to "still-hunt," or stalk a buffalo, if they could find one,
to view the country more to the north, and to ascertain
its promise of game. Major Martin, Willie, and myself
kept each other company, at times when I was not occupied with my notes, until the afternoon, when taking
Brutus and one of my John Manton double-guns, loaded
with shot and cartridge for wolf, deer, or game, I
strolled up the wooded sides of the creek, thinking that
after so stormy a night some of those wolves who were
so perpetually howling in our vicinity would probably
have curled themselves up to sleep in the sheltered,
grassy, and bushy sides of the high banks above the
stream. The creek was so narrow, that with my powerful Manton I could command it all. Its recesses, however, though closely explored, were a blank, and Brutus
tried the cover all in vain.

Having taken quite exercise enough to keep myself in
condition, I then returned to camp, where, as it began
to get dark, Major Martin expressed some uneasiness for
Bayard's return, and we kept a look-out for him, orders
being issued that after sunset an occasional shot should

be fired to direct the hunter on his return, in case he had lost his way. There was this additional cause to be sedulous for Bayard's safety, in that, supposing the Indians to have somehow or other got at the fact that Bayard had slain Pawnee, the brother of Tchorson, there was not a "brave" in the whole tribe of that chief, the "Kiowas," who would not gladly have risked his life to have revenged the death of the petty chief, and possessed himself of the scalp of the white warrior. We were all well alive to this, and knew that, had the Indians been aware of the position of things, we should to a certainty have had "Setanka," the father-in-law of the late Pawnee, and the war-chief of the Kiowas, upon our camp with a demand, for the purchase of our own safety, that Bayard should have been given up. Of course this would have brought on a battle, and "a very pretty quarrel" it would have been! Immense odds in numbers on the side of the redmen, but arms, ammunition, and the best shooting all along with us. To assign to the Indians the unerring aim with the rifle awarded by novelists, is nonsense; the savages dread the weapon, as well as the better shooting of the white man, and take very good care not to expose themselves on anything like equal terms, in point of situation. At from twenty to thirty yards their bows are a nastily-effective weapon, for they shoot correctly and hard enough at that distance to kill, and the wounds they make are difficult to cure. The way the bow and arrow tells most efficiently is by night, if there is cover enough to conceal the assailing foe; there being no flash and very little noise to direct the rifle-fire in reply, it is of course difficult to quell the unseen annoyance. On such occasions as these, how

well blue-lights would tell! and, were I to go again to
the plains, I would never be without them.

The night had not long set in when Bayard and Mr
Canterall arrived, so with hearts at rest we saw him sit
down to a comfortable dinner. Before reaching the
plains all the Americans were ringing in my ears the
deliciousness of the buffalo meat. " Reckon, sir, you've
nothing like it in the old country. Yes, sir, just about
a treat you air going to have ; yes, sir, you'll have something *to* speak on, when you go back, that's a fact."
When I came to test the matter, I found that it was the
" hump rib, " *not the hump itself*, that was the best part of
the bison, and the meat along and on either side the
loin, the tongue, and oh, shade of Eude, the marrow-
bones! No man can guess what marrow amounts to
until he has been to the Far West and eaten it as Wallace, who cooked on the plains for me, dressed it. The
bone was brought to table in its full length, and they
had some way of hitting it with the back of an axe
which opened one side of it only, like the lid of a box.
The bone then, when this lid was removed, exposed in
its entire length a regular white roll of unbroken marrow, beautifully done. When hot, as the lid had kept it,
and put on thin toast, it was perfection! On inquiry I
found that the two extreme ends of the marrow-bone
only were placed on the red embers, and the heat of the
bone itself dressed the marrow. As far as the bison
meat went, it was precisely lean beef, with no more
flavour than lean beef in England would have ; while at
the same time, as we were living from hand to mouth,
and had no facility for keeping things, of course it was
always tough. The veal was precisely lean veal, such as

we have when unfattened in England. The beef of a
bison heifer I never tasted, as we did not kill one; but I
should imagine that that, in regard to comparison with
ox beef as fattened in England, or as to our cow beef,
was as inferior as that of the bison bull to the ox, but
no doubt the bison heifer affords the best food on the
plains.

On the morning of the 12th of October we struck camp
in very doubtful weather, and continued our march to the
west in search of buffalo, but ere we had proceeded far it
came on to rain again, and looked very like a wet day.
The plains were in a terrible state for the waggons, so
taking counsel together, we countermanded the line of
march, and ordered the waggons back to the same creek
whence we started, directing Mr Canterall to return with
them and select a more sheltered spot, and a better one
for wood and water, but to outspan sufficiently near the
same place, to enable Bayard and myself, who were to
proceed to stalk buffalo, to have no difficulty in finding
the camp, in case we should have been overtaken by the
night.   Major Martin returned with the camp in command
of all.   Having lent Bayard Pape's double breech-loading
rifle, and taken with me my old powerful muzzle-loading
rifle, by John Manton, as well as Prince's breech-loading
carbine, and Phill, on my pony, to hold our horses and
carry one of my rifles, we started, Bayard on a hack horse
and myself on Sylph.

We had not long parted from Mr Canterall and the
waggons ere we saw that rascal follow us over a hill and
make a signal with his hat towards a herd of buffalo feeding
on a hill, which we had already seen.   I made a sign to
him with my hand that it was all right, and that he might
return to his duty, and then Bayard and myself made a

detour to avail ourselves of the wind and some low ground.
Having reached the shoulder of a low hill, which again
gave us an available line, Bayard suddenly reined in his
horse, and, pointing with his hand, called on me "to look."
At first I could scarcely believe my eyes, and so doubtful
was I if they served me right that I exclaimed to my
companion, "Why, *is it* that fellow Canterall?" "No one
else," replied Bayard, "and he means to have a shy at our
game!" I believe that if it had not been for the fear of
hitting my own horse Kansas, whom he had been strictly
ordered by me to keep quiet, I should have fired a long
shot at this rascal to have stopped him; but resolving to
see what it was he was going to do, we contented ourselves, unseen by him, with observing his actions. We
saw him, on my steady horse Kansas, run at the herd;
he was beautifully carried up to them, and we heard him
fire five or six shots, and knew from the way he went
that he had disturbed the whole line of the country over
which we had resolved to go. We therefore at once
struck off in a much wider and more easterly direction,
and saw nothing more of him, and it was very well for
him that we did not just then do so. The rascal in this
instance, as usual, missed everything he shot at from his
horse, except the calf and a wounded bull, of which I have
before spoken, and on this occasion spoilt our sport and
disobeyed every order he had received.

Bayard and myself continued our way over the plains,
and rode right into a large pack of prairie grouse, which
lay very well, and longed for our shot guns; but not
having them, we kept after the largest game. As we
were riding together, I fixed my eyes on something just
the colour of the stem of a large tree protruding from
some very high grass, and called Bayard's attention to it.

It was motionless, and a long way off. It could not be
the stem or root of a large tree, because there were no
trees on that part of the plain; then what could it be?
We stood on a gently rising eminence narrowly watching
this odd-looking object, when suddenly from the grass
immediately under it a flock of black birds arose, about
the size of the English starling, and in habits and flight
very like it, and Bayard's glance instantly brightened
as he said, " Buffalo—it must be a buffalo, for there are
the birds that are always about him." The buffalo, or
whatever it was, never moved, and the flock of birds again
vanished in the grass.

Having dismounted, with rifle and carbine under each
of my arms, our horses delivered over to the care of Phill,
who was only to advance to us if we made a sign for him
to do so, Bayard and myself then availed ourselves of
wind and low ground, and began to creep to this suspicious-looking excrescence on what seemed to be the
level plain. Having approached some distance nearer to
the object, we crawled on our hands and knees and peeped
over a little hillock. There was no doubt about it then,
for the excrescence we had seen was no other than a buffalo's hump, and on the same spot there stood the owner
of the hump, now fully confessed, in the shape of a monstrous old bull. We backed ourselves down the little hill
again, having taken a needful glance at the ground, and
on our hands and knees re-commenced a covert advance.
Having my rifle and carbine in my hands, this hand-and-
knee work was difficult. Again we came to a little rise
in the prairie, which would enable us to make another
observation, as from the very slight undulations of the
ground and variation of the low parts of it the stalk had
to be governed by the greatest caution, for when on ·

our hands and knees the grass did not reach our backs.
On attaining this next spot of observation I put down my
guns, and looked over the little hill, and to my astonishment, instead of one, there were two huge bulls standing
very quietly together. Having made every needful
observation, I returned to Bayard, told him what I had
seen, and we again got into the little hollow, well knowing that, when we came to another rise in the plain, when
we looked over that we should be within eighty yards of
the two bulls. On reaching this spot, we both peeped
through the grass, when instead of two old bulls there
were now three, all standing quietly together, and the
ground afforded us even a further advance well screened
from observation.

We reached then the last spot to which we could attain,
about fifty yards from the game, when, having first ascertained that the three old bulls were quite quiet, we
lay down to rest a moment from our long crawl on hand
and knee, to steady our hands, and to gain our breath.
We then agreed that I should take the first shot with
my single Manton rifle at one bull, and that Bayard
should then fire at a second; my carbine and his
second barrel to be used according to circumstances. An
agreement was also made to fall on our faces the moment
we had fired, if either bull was then and there disabled,
to avoid being seen by an animal rendered furious, who,
on such occasions, almost to a certainty, if he detected
his assailants, would charge and trample them to death,
unless killed by a reserved fire. I then took steady aim
at the large bull that offered the fairest broadside; but,
though I had purposely studied the figures of the bisons
at Taymouth Castle until my long supervision of them
made them give me a hint to be off, from the height of

the hump, as well as the elbow and brisket being concealed by the unusually long grass, I hit the bison too
high, which was immediately evident by a gush of blood
about mid-way in the body, and if in a right line for the
heart, still much too high.   To our astonishment, the instant my rifle was off, a fourth bull, still larger than the
other three, seemed to rise up out of the earth, and Bayard fired and struck him, but too much behind, as the bull
was going obliquely away.   Their backs being all to me
in flight when I caught up my carbine, I reserved the
fire, and made a sign for our horses.   On their arriving
I handed my Manton rifle to Phill, and gave Sylph the
office for a run, which she was too happy to obey.

I soon found out, ere I had gone fifty yards, that these
four old bulls had been down in a small rill of water,
which was very deep, among high grass, and not perceptible to us from a distance ; and this was the reason for
three of these bisons being at one time completely out of
sight, and for the hump of the fourth alone appearing.
Sylph was soon over the water-course, and coming up
hand-over-hand with her game, when, before I ran alongside, I saw that the bull I had stricken in the right side
—the heavy round ball having made a most unmistakeable wound—was dropping behind the others, and that
the immense beast, the king of them all, who got out of
the water on the first fire, and whom Bayard had hit,
was doing the same.   These two wounded bulls kept at
about the same pace close together and nearly side by
side, the one I had wounded being about a head in
advance.   Sylph, keeping them on the left, went up
to them well, though slightly swerving and shaking her
head and shoulders on the rising of the carbine to fire,
and I shot at the bull I had previously wounded ; but,

missing him, the bullet took effect in the hock of Bayard's bull, and in a few more strides the monster stopped for mischief. That beautifully handy carbine was soon re-charged, and the next bullet went close behind the elbow, and again the same, and the gigantic game fell dead.

When he was down, Bayard and myself stood over him in admiration. He was much the finest and largest animal I had seen since I had been on the prairies; but Bayard, who of course was a better judge than I was, said he was the largest bull he had ever seen in his life. His robe was beautifully grown in silky young curls from hump to hock, while his mane, beard, and shaggy fell of young black hair from withers to dewlap, was the most luxurious thing that could be imagined. We measured him on the spot as far as we could, and he was upwards of 5 feet 8 inches from top of hoof to the top of the hump, and 9 feet 6½ inches from his nose to the root of his tail. The bone of his fore leg, between fetlock and knee, was 11 inches in circumference. We were unable to measure his girth. So beautiful was the robe of this monarch of the Far West plains, that we resolved not to spoil it by taking the meat, but if pos-sible to send for it from camp. We then only took the largest tongue we had ever seen, and the tail, and left him, alas, for the wolves; for when we reached camp it was so late, and the distance to this buffalo so far, that it was quite out of the question to attempt to recover any further spoils. Mounting our horses, and following in the line of the other wounded bull, an excla-mation from Bayard drew my attention to a black mass lying in the grass before us, and we together exclaimed, " Ah! ha! then, here he is!" On reaching that to which our attention had been called, there,

indeed, lay the carcase of a bull, but not the one I had
shot at, for it had been dead a couple of days, and had
been in part feasted on by wolves. We continued on the
line of the stricken game till we attained some high ground
which commanded many miles of country, across which
their line of flight appeared to have been directed, but
not a vestige of game being to be seen, we gave up the
chase, and proceeded to reconnoitre undisturbed ground.
After proceeding a considerable distance, we opened out
a very wide stretch of the plains, and I never saw a more
curious and beautiful sight than was then offered to my
view.

The disturbance among the bisons we had hitherto
occassioned, either by the flight of alarmed animals or by
the noise of our rifles, had certainly not preceded us
thus far, for the entire vista for miles on miles was a scene
of rest; bisons were dotted about singly as well as in
small herds, either lying down or quietly browsing on the
grass, and we had only to select the game that was the
most easy of approach. On prairies such as these, the
low water-course, dried but grown up with rank weedy
grass, provided the direction of the wind will suit, offers
the best and almost only approach; when, after closely
scanning the country, we determined on a stalk at five
large bulls who were feeding within shot of the dry but
very shallow ravine. We had doubts when we left our
horses and began the stalk if the wind would always
avail us, but, on reaching the low ground, to our great
joy, we discovered that a branch of the dry water-course
would enable us always to be down wind of the game till
we came to a certain distance, and then we might select
our own course in the grass. When we dismounted, we
were a considerable distance from the bisons, but we

desired not to attract attention by the display of horses, because at first we could not be sure if the stalk we were thus commencing would be throughout available.

Having crawled an immense distance on our hands and knees, or, when the depth of the cover would permit of it, in a stooping position, we began to near the huge bulls, when on peeping through the grass we saw that one of them had singled himself out, and that, could we but leave the water-course, and conceal our further advance sufficiently in the low prairie grass, we could attain to within eighty yards of him. Having then reached the extreme point where the cover of the ravine was available, we took a long breath to prepare ourselves for a crawl on hand, knee, and breast, and again advanced. We crept on thus till the grass got so thin that we dared go no further, when, rising slightly to our knees, we fired at the bull as he stood sideways to us, knew that we had hit him, and then fell flat on our faces. Having given the smoke of our rifles time to subside, we peeped through the grass, and saw that four of the bulls were staring about them, not in the least aware of what the noise was ; while the one we shot at had walked a little distance away and lain down. We knew, therefore, that he was severely wounded, and in all probability as vicious as an animal of that size had it in his power to be, if he could only find an assailant on whom to wreak his vengeance. He had attained to perhaps more than the distance of a hundred yards before he had laid down, and, when down, he had assumed a position that gave us in our present situation a very bad offer of a shot, for he had his head away from us, and his side therefore not available. The grass, too, was so very scanty, that it did not admit of a change in our position without the certainty of being seen. The four companions of this bull then walked quietly away, but he showed no disposition to follow them. On this we held a consultation as to whether we would advance for another shot, and risk his coming at us, or whether Bayard should return and bring up our horses, and thus make more certain not only of our own safety, but of the game. I ruled that we should have our horses; therefore Bayard went back to fetch them, while I remained lying in the grass to watch the foe. On observing that he showed no disposition to move, I also crept back to the water-course, where I could stand up sufficiently to load my Manton rifle, and in a little time Bayard returned with the steeds, and we ordered Phill to lead them in our rear, and to bring them up in an instant if we gave him the sign. On crawling to the spot whence we had fired, the game lay still in the same position; so, making a slight detour, at the risk of being seen, to have him in a more available position, we again fired and fell on our faces for an instant, when, looking up, we saw him, though again stricken by both balls, set off in a long trot over the prairies after his companions. A sign to Phill and he came up with the horses, when I mounted my pony, wishing to give Sylph a rest, and Bayard and myself set off in pursuit.

As I expected, the bull was too severely hit to go far; he broke at first from a trot into a canter, but on finding himself pursued and overtaken, on coming to a rill of water surrounded by high reeds he stopped at once on the edge of it, and, in a threatening attitude for a charge, stared wildly on us both. At that instant we gave him three more balls, when, blundering forwards, he fell headlong into the stream, which he completely bridged with his body. Having delivered our horses to Phill, we went

SOMETHING IN THE BUSH

to the bull, who, as Bayard passed over his body to get to the other side of the stream, as the beast lay right across it, shook his head at him in a dying menace, and then ceased to exist. I was kneeling on the bison's shoulder, and holding up the head for Bayard to get at the root of the tongue, when my companion suddenly dropped his knife, stood erect, and laying his hand on the pistol at his belt, stared curiously into the high reeds behind him. To step back to the other bank and pick up my carbine and to rejoin Bayard was but the work of an instant, when I asked " What it was ?" " Something in the reeds," was the rejoinder ; when together we left the body of the bison and inspected the cover. There was nothing there, however ; so, having made quite certain that there were no lurking Indians near us, we returned at our leisure, possessed ourselves of the tongue, loin steak, and tail of the game, and proceeded to tie it to our saddles. While doing this the rustling in the reeds which had drawn Bayard's attention was explained, for two beautifully dark wild faces with sharp pricked ears suddenly rose out of the grass, not fifty yards from us, disappeared, and then rose higher still and further off. Two large wolves no doubt had scented the game, which must have fallen almost upon them, and were immensely curious to ascertain if we were going to leave it. I snatched up my carbine and mounted Sylph, to see if they would let me come nearer, for they had again disappeared in the grass. They did show on their hinder legs once more, at about 150 yards' distance, and I fired from the saddle, but without effect. We then set off home, and were rather puzzled to find the new position taken up by the waggons, for the smoke did not rise, and our attention was rather drawn

off, from our mistaking at a great distance some " shade
bushes," which had been stuck up in an old encampment
by the Indians, for our picketed mules ; however, just at
dark we made out our camp, very hungry and rather tired,
and longing for some dinner.

On arriving we found that Major Martin and Willie had
gone out after some bisons near the camp, with Mr Canterall, who took his rifle for a " still-hunt," or stalk.   He
succeeded in wounding two, at whom he fired forty shots,
as counted by Major Martin, ere he brought them down.
George Bromfield had also given his American companions a sample of an English New Forest shot, for with his
rifle he killed, hitting the bird right through the breast, a
single wood-duck who flew over his head.   I had never
before seen this beautiful bird in his native state, and
when dressed for dinner the next day he was delicious.

" Well, Mr Canterall," I then said, " what became of
the herd of buffalo you signalled to us with your hat
soon after we left you ? "

" Oh, they got frightened at the passing of the waggons, guv'nor, that's a fact, and moved off."

" Why, did not you chase them ? "

" No, guv'nor, I didn't leave the waggons."

" What a vile untruth ! " I cried ; " why, Mr Bayard
and myself watched you run them and shoot at them."

This fellow, as he always did, replied in a whining
apology, that he was very sorry if he had been wrong,
&c., &c., and I turned my back on him in disgust and
went to my tent.   My mare, Sylph, had not recovered
her previous hard day, and, in addition, she had caught
cold in the terrible rain of a night or two before.   Had
she been fit to have gone the chase, I could have passed
the great bull we killed, after wounding him sufficiently

in the hock, and have recovered the one I had stricken too high in the body with my first shot; but under the circumstances, we were quite satisfied with our day's sport. As a further proof of the insubordination of the men that I had hired, one night, when Major Martin went the round of the camp, he found the same man that I had already spoken to neglecting his duty at his watch, and when Major Martin reproached him for it the man was again very insolent, but which insolence Major Martin again cut short by threatening immediate expulsion from the camp. I give these two instances of neglect, when most attention was required, to show what little dependence could be placed on such men if really in the presence of an enemy or of the most expert animal thieves that the world affords. Subsequently I was much pleased with the same offender, Tom, and I believe that in matters he liked, or in mere sporting details, he would have made a very useful follower, and a much better guide, not only across the plains, but to the haunts of game, than the useless and false fellow so improperly placed at the head of my travelling establishment.

I regret to say we could not send for the robe of the mighty prairie king whose dimensions we had taken; but all sensations of lassitude from fever having now left me, I felt well up to any amount of work that the prairies might yet hold in store.

CHAPTER XVII.

THE next morning, the 13th of October, the weather
was beautiful, when, mounting Taymouth, and armed
with my Manton muzzle-loading rifle and Prince's carbine, with George Bromfield on my pony to carry the
Manton rifle and to hold our horses, Mr Bayard and
myself, in company with Major Martin, set out for some
rocky plains, unsuitable for running, for the purpose of
stalking buffalo. Mr Bayard again had my double
breech-loading rifle made by Pape. As this was to be
the last day entirely devoted to buffalo, we took a six-
mule waggon, Mr Canterall, forbidden by me to hunt,
and five men, in order to possess ourselves of such spoils
as the chase afforded. We were proceeding over the
plains, on our way to what we conceived to be undisturbed ground, when suddenly a wild-looking single bull

buffalo came trotting over the brow of a distant hill, and
Mr Bayard and myself at once guessed that he must
have been chased by man. The mystery was very soon
set at rest, for after a little while two human figures on
foot appeared over the same brow, as if running in chase,
when, having fixed his eyes on them for a moment,
Bayard pronounced them Indians, and exclaimed " Will
you come ? "

Of course the reply was in the affirmative, and we set
off towards them, Mr Canterall in company, when as we
thus proceeded I asked Bayard what he was going to do ?
The reply was, " To see what Indians they are; and,
if enemies at war with the United States, to war on
them." " All right, but how shall we know what Indians they are ? " " If they are of the tribes at war, on
seeing us they will fight or run," replied Bayard; " if not
at war, they will wait our arrival for a talk." I was
charmed thus to have fallen in with a chance for higher
game than even the bison afforded, and we kept on towards the men at a fast but collected pace. That they
were armed with rifles I saw by the gleam of the sun on
the barrels; but I had not gone very far before I exclaimed to Bayard that I doubted if they were Indians;
but as his eyes were younger than mine, as well as more
accustomed to the plains and the life upon them, of course
I deemed him the better judge. He replied, " They are
Indians," and we kept on, when they rather turned to
meet us, and by a nearer approach soon set the point at
rest; they were two white men, if they had been clean,
but their faces and hands were so disguised by hair and
daubed and begrimed with dust, that until we were
close upon them it was almost impossible to declare their
hue or nation. We greeted each other, and they then in-

formed us they had only been out a few days, had seen
no Indians, and had killed two buffaloes, concluding their
information with a moderate request to me to ride after
the buffalo they had been in pursuit of, then at least two
miles off, and head him back. This I respectfully declined, when, on our road back to the waggon, Bayard
told me that the thing in the extreme distance that deceived him was the whiter (certainly not white) shirt of
one of the men, which having found the light very considerably from a rent in his garment, it at a distance resembled the breechclout of the red man, and induced him
to make the mistake.

Soon after this we sighted another single wild-looking
and unsettled bull, no doubt one of the herd that these men
had been after, and we tried to stalk him, the ground being unfavourable for running, but in vain. Having continued our search more to the westward, we then came to
a portion of the desert that was in profound peace, and
we could see herds of bisons in different directions, but
none of them available to us in position or in regard to
the wind. There was a small herd of bull bisons, however, that appeared to be feeding in a place up to which
a dry water-course with long grass in it seemed to lead.
Though that was not quite as we could wish in the then
state of the air, still it was the best chance that offered ;
so, giving our horses to George, we began a cautious approach to the water-course over some space of the plain
commanded, as we feared, by two of the bulls, if by
chance they ceased from grazing and happened to look in
our direction. We succeeded in crawling over this open
space, lying flat on the earth and quite still when a head
of the game seemed likely to be lifted up, and when once
in the high grass of the ravine we felt in comparative ease.

Along this water-course, then, we went for a considerable distance, stooping very low, when suddenly, to our
great chagrin, we found that in one of its detours it would
give the wind of us to the bisons instead of our having
their wind, and we came to a halt for deliberation. On
taking a very searching view of the land, we made up
our minds that the water-course deviated into a sort of
elbow, and that if we could cut off the bend, and get into
its course again, we should avoid giving our wind to the
game. Once more, then, we had to crawl with all possible caution through the short grass of the higher land;
again we succeeded in our endeavour, and found ourselves
a second time ambushed in the long grass. We were now
beautifully sheltered, as well as completely favoured by the
wind, when, arriving as we supposed within shot of the
buffalo, we ventured to look through the grass, but found
that we had not justly computed the distance. The bull
nearest to us was still too far off, but feeding towards us
and away from his companions, and absolutely approaching the ravine, along the course of which we were coming. The next time we halted to make a reconnaissance,
and before we were within shot, the immense beast came
up to the brink of the ravine, stepped into it, and, with
his head still in the grass at feed, absolutely progressed,
though very slowly, full on the line to meet us. Every
time we stopped for a few moments of rest and observation we could see, through the reeds and grass, the huge,
russet-coloured, conically-shaped hill of flesh right in front
of us, and evidently, but very gradually, coming directly
for the spot on which we lay. Occasionally the ferocious-
looking head of the bull would be lifted, chewing a
mouthful of grass, and while the jaws were so engaged
the ruffled and shaggy black face of the monster would

appear to be looking straight at us. When the head of the beast was thus up we lay without motion, but when he lowered his head we crawled gently on again to meet him.

The next time we paused for rest and a view of the game, we found ourselves close upon the bull, certainly within forty yards of the great and gently moving mass of hair, which, when the head was lowered, looked in shape like a gigantic bee-hive. Still and still the bull came stem on, as a sailor would say, and still he offered to our rifles nothing but, as it were, woolsacks of hair, or the top of his shoulders and his hump. Bayard insisted that it would be useless to shoot at his forehead, for that the hair there was in such a matted and a tangled mass as to be ball proof against any rifle that ever was made. We were thus obliged to lie flat on our faces, and flat as we could lie, from our chins we now saw the top of the hump, looming more largely into view as the beast came on upon us. "He'll tread on us soon," I whispered to Bayard, "and we shall have to fire up his nose!" and when I looked at the bull again, and then at myself and Bayard, I could not help thinking what mere frogs we were in the grass when compared to our giant foe. And supposing he took it into his head at once to charge, what would then become of us? I had scarcely made this last observation, when his companion bison, some hundred yards' distance, who had observed the herd to which they belonged moving away, walked off, and our game lifted his head, not above twenty yards from us, to look at him. "He's going to turn," whispered Bayard; and accordingly the bison did turn, with an evident intention of walking after his companion, when at that moment, and with steady aim, our rifles were fired, and then together we fell flat

Printed by]

Spottiswoode & Co.

STALKING THE BUFFALO.

upon our faces. We had not the least doubt but that we
had wounded him mortally—my aim as well as my ear
assured me of that fact—when, having given a moment's
space, in order that if the bull had looked towards the
position of the noise he might have satisfied himself that
no enemy was there to be seen, we both raised our fore-
heads sufficiently to observe that the hump was moving
slowly away to our right, and then gradually it disappeared in a fashion to indicate lying down rather than a direct fall. The monster was still not much above forty yards
from us, and very probably, if not dead, as furiously savage
as a mortally stricken beast of the size and age could be.

And now became manifest the great superiority, in
situations such as this, of the breech-loader over the
muzzle-loader. I dared not, for the life of me, kneel up
to load my John Manton rifle, or I should have been confessed to the savage and dying foe. Had I done so the
bull must have heard and seen me, and had he charged
we should have been dependant on three shots at his
said to be impenetrable head. Bayard, laying flat on the
ground, charged Pape's breech-loading rifle in a moment,
and without the necessity of drawing the slightest attention from the wounded beast, and I had still Prince's
breech-loading carbine. As soon as he had reloaded we
again crawled towards our foe, when, on looking through
the high grass, we saw him lying down, and looming as
wildly savage as those very savage-looking animals can
look—his head and body obliquely away from our position.
It was no use to shoot at him thus, so we resolved to
take to the short grass, risk being seen, and open out his
broadside, determining that if he detected us, and rose
to charge, our fire should be at his heart while in the act
of turning, and Bayard's last barrel retained in the event

of coming to close quarters. We crawled till we had
opened his side, and he never either stirred nor saw us;
so directing our balls quite low as he lay, we fired together, when the monster sprang to his feet and stared full
upon the spot where we lay. There was then a most
anxious but beautiful pause, when, on seeing that the bull
had not the least idea of our close proximity, but that he
began staring over us and towards the ground on the
other side the ravine, I whispered to Bayard to back into
the long grass, or he might walk on till he trod upon us,
and that when there we could reload. This we did, till
we began to find the descending ground, and then we
turned and slid uncommonly fast on our waistcoats until
we were safe under cover. When there, we lay for a few
seconds convulsed with laughter at our own haste to get
back out of sight, and then re-charged our rifles. The
moment we had loaded we crawled back to the short
grass, and took a look at the position of our foe. The
least rise of our heads ought to have afforded us a sight
of him had he remained on his legs; but no, so higher
and higher went our brows, till at last we saw the monster
extended on the plain upon his side with his head towards
us, when, on attaining to our feet, we found that he was dead.
A signal to George with the horses, again transmitted by
him to the waggon, brought up our people; so, leaving
them to take the robe as well as other spoils, Bayard,
myself, and Major Martin again set off in search of other
adventures, indicating our probable course to Mr Canterall, and desiring the waggon, if possible, to keep us in
sight by holding the hills, and always to maintain such
commanding ground as would enable the men to view us
on the plains and to be aware of the course we were taking.

Ere we departed, an inspection of the carcase of the
bull gave good evidence of the exactitude with which we
had shot—the four balls had gone in very near each other,
and close behind the shoulder, the only fault being that
they were still a little too high. I account for this not
only in our not having sufficiently allowed for any little
rise in the projection of the balls, but also from the immense depth of the animal, made still more deceiving by
the height of the shoulder and hump. This buffalo had
no tail; there was a little stump where the tail should
have been, not an inch long. Perhaps, when he was a
calf, a wolf, in an attempt to catch him, had bitten it off.
Leaving our men occupied in skinning the bison, we then
went onwards, and at a long distance from us beheld a large
herd of buffaloes slowly going down to cross the creek on
which we were encamped. Their intention was evident,
and we communed with ourselves if it were possible for
us to reach the creek first, and, under cover, attain the
spot at which they would pass; but on computing the
distance we should have to go to get out of their sight
and under the banks before them, we found the thing to
be impossible, so we contented ourselves with watching
them. They crossed the creek, and then went feeding
over the plains very leisurely on the other side. Letting
them get so far beyond us that we also could be over
the creek before they took any alarm, we then surmounted the difficulty, and saw that two or three buffaloes were in a position to be stalked, and we dismounted to try to reach them; but they saw us, and rejoined
the herd, which had fed on to very rideable prairies, where
there was no chance of getting near enough to reach them
save by a run on horseback, and on that we mutually resolved. "Now! my dear old horse," I whispered to Tay-

mouth, patting his neck, "this is to be the last arranged day for buffalo running on which I ride you, and you had better be quiet; I don't mind how much I take out of you, but a bison you must kill!"

Long before we got near the herd, they saw us, and all set off in their remarkable up-and-down canter. The last thing I knew of Bayard, until the chase was over, was, that he started when I did; Taymouth then flew over the plains, and I was soon alongside of the bisons and saw nothing else. No sooner was I up with them than Taymouth swerved away, and when I checked him he began fighting in that usual changing-leg way of his, which rendered it, in his violent moments, so difficult to take any aim from his back. I got him as near as I could to some immense bulls, and fired and missed one; reloaded and fired again, but with similar effect; reloaded a third time, very angry with my fractious horse, and shot again, and this time, I think, with some slight effect. There were two fine old bulls going side by side, at whom I fired the third shot, and I suspect that the ball must have struck one of their horns, for with a sudden start they both struck off from the herd and went away by themselves. "Now then," I cried to my horse, "with both or one of these huge bulls you shall go till I quiet you or a bison;" and at them I rode, but only at both for a few strides, for now they separated, one inclining to the direction of the herd, but the other, the largest of the two (as fine a specimen of the giant game as could be seen), kept straight away; and this was the one I selected, on to whose traces, whithersoever he might go, I resolved to attach my fractious horse. Thus, then, we started—the bull gaining a considerable space while I pulled in sufficiently to reload my carbine; but when that was loaded, I but shook the snaffle-rein, just touching the curb enough to arch the horse's neck and induce him to see where he was going, when Taymouth flew after the bull, and came up with him hand over hand.

The bull knew very well that I was pursuing him, for no sooner was he aware of my approach than he turned short to the right at rather less speed, evidently furious to charge if he got an opportunity, and made at once for a small branch creek that ran into the one we had so lately crossed. On seeing his object, I pressed to within thirty yards of him, in order to be on the bank he left, for a slower shot at him as he ascended the opposite one ; and observed more surely by his leer at me from under either horn that if he had but a chance he meant mischief. Quieting Taymouth as well as I could, to induce him to stand still, the instant the game began to climb the opposite bank I presented my carbine and fired, but the unsteadiness of my horse as usual interrupted my aim, and again the buffalo made play upon the plain, evidently hoping that he had shut out his pursuer. Having reloaded, Taymouth slipped down the bank and leaped the boggy water, sprung up to the plains again on the other side, and again took up the running. Precisely the same thing happened again—the instant the bull found himself overhauled he turned short to the right and made for another branch creek infinitely larger than the last ; and as I pressed close upon him for the same purpose as before, I saw him, at a gallop and without the smallest pause, go right over the edge of the high bank, and disappear from my view.

On gaining the exact spot where he disappeared, I perceived the creek was very wide and swampy, and full of high reeds, but from my commanding position I ought to

have seen the bull, turn which way he would; there was,
however, no such animal in view! "East, west, alas! I
knew not whither" the bull had gone. He had not had
time to ascend the opposite bank, and if he had, the
prairie was open and I must have seen him, and he certainly was not going in the hollow to the right or left, or
he would have been equally visible. There was, however,
a short turn in the creek, and it was just possible, though
I much doubted it, that he might have reached it, and so
have shut me out of view; I therefore let Taymouth feel
the calf of my leg, and shook the snaffle, for the purpose
of following the bull, at the exact spot where he had so
unaccountably disappeared. Taymouth, however, to my
surprise, refused to go; when, still with my eyes directed
to the creek and plains beyond, expecting to see the
monster reappear, I put my horse at it again, but he turned round more resolutely than ever, and began to jib and
rear. This struck me as very odd, because the difficulty
in the chase with this hot chestnut horse was to prevent
him from rushing blindly at anything that looked like an
impediment to headlong speed. My curiosity was therefore aroused, and when I had overcome his restiveness,
and put him at the place again, I looked at the same time
to see of what it was that he was afraid, when, O blessed
St Hubert! close beneath me, on the very spot where he
had disappeared, there lay my monstrous game, couched
in the reeds on all fours, and hoping to find me and my
horse come down upon his back, and so deliver myself
to his power.

The bank was very steep, and the bottom boggy where
the beast lay, and I could see his ears moving and listening up in expectation to catch any noise I made. Oh!
what a shot his spine and broad back offered, right be-

neath me, and not much more than seven yards below!
Sideling Taymouth to the edge of the bank, I raised my
carbine for a shot, but the instant that was done a slight
swerve from my horse shut the bison from my view, and,
O fabled fiends of the Hartz mountains! Taymouth would
not let me fire! It made me desperately angry this, but
rage availed not, so, after trying to induce my horse to
permit me to fire, but all in vain, I dismounted, under
the idea that if my friend below charged up again from
his sort of saw-pit for the duel, I could mount in time to
prevent collision, though in the fidgetty state of Taymouth
it might have been a very near thing.   Patting my
horse's neck, and trying to quiet him, with the rein over
my elbow I then again stepped to the brink of the ravine;
but the instant the carbine was raised my horse shied
and pulled me back; and again and again I was baulked.
It was a sad dilemma to be in, but there was no help for
it, save by pacifying my steed, and this I attempted still
to do by caresses, and at last I induced him to stand
directly over the buffalo, on whom he kept his eye, provided I did not attempt to fire; however, hoping that by
degrees I should succeed, for a minute or two I continued
to do no more than pat and kiss him, my eyes looking
down on the foe below, who lay quite still, all but his
ears.

The first symptom of greater quietude was in my
horse rubbing his head against me; the next was, he
heaved a deep-drawn sigh, as strong-lunged horses will
do on regaining their wind after a severe gallop; but,
alas! his returning self-possession and recovery of his
wind did not end there, for having caught his breath,
he blew his nostrils with such a healthful and emphatic
snort that he might have been heard half a mile off; and

the noise being new to the buffalo, the bull shot up and out of the reeds in which he had been lying, like a bolt from a bow, and sped away from me across the ravine as fast as his great strong gallop could carry him. As the bull jumped so suddenly up, making such suction noises with his immense hind legs in the bog, Taymouth flew back and twisted round to such an extent, that he nearly tied me up in a knot. However, I disentangled the reins, and jumped on him in time to take a random and much-interrupted, and therefore ineffective, shot at the bison as he climbed the opposite bank, when, not wishing to lose sight of him, I put Taymouth at the ravine, into which, now that his enemy was gone, he was but too happy to descend; and, as we came to the couch of my retreating friend, we were considerably bogged, but not so much but what my horse beautifully and safely plunged through. Having also attained the opposite bank, and caught a view of the bison going over the plain, I slackened speed, reloaded, and again took up the running, letting my horse feel the spur and go at full speed, by way of quieting him against the next event. No sooner did I begin to come up to close quarters than the buffalo as usual slackened speed, and with a sulky leer at me from beneath his horns collected himself, and, turning from the open plains again short to his right, he ran straight for the great main creek, into which the smaller ravines which we had crossed ran. I pressed him closely, and when he went as usual slap over the bank into the large creek, I might have been about thirty yards behind him; and by the water which flew up as he so suddenly went down, I knew that we were close to, or in, the water dammed up in places by the beavers, which I had previously seen.

This main creek was not so wide as the last, but much deeper, the sides excessively perpendicular, and the water I found, on sliding down the bank into it, about up to my horse's girths. The bison had begun to clamber up the other bank when I reached the water, so dropping the rein to my horse, in the hope that he would occupy himself with drinking or pawing, and enable me to make a successful shot, I brought my carbine to the shoulder; the instant I did so, however, Taymouth, who would do nothing else than keep his eyes on me and the retreating foe, flew right round as the rifle came to the level; and while I again missed my shot, he, at the same time, charged up the bank down which we had descended; and, alas! I dared not check him till we had reached the top, or we should very likely have fallen backwards. He then got both spurs furiously into his sides, and we went again into the creek, and as I set his head at the opposing bank he had the spurs all the time, flying like a bird to the very top, though it was so steep that even the buffalo had had a considerable scramble to surmount it, and when on the plain again my horse was mad to pursue his foe. I slackened speed, however, to load my carbine, and then took a view of the scene before me, as I set off after the game. The buffalo was making the most of his time, in apparently perfect knowledge that he had put a large difficulty in the way of his enemy, which might occasion very considerable delay, if not altogether prove the means of complete escape. His head was down, and he was racing at the very utmost of his speed, but oh! splendid sight, right in front of him, and for which land of promise he was so swiftly proceeding, lay prairies of from fifteen to twenty miles in extent, without so much

19

as a bush or ravine of any consequence, or iron-faced hill
to stand between the mighty bison and myself, or to im-
pede a close encounter.

"Now then, Taymouth, the time has arrived when
either you or the bison, or both, will be quieted;" and,
thinking thus, as I patted the swelling veins on the arched
neck of my horse, I drew his reins the tighter, not only
to husband his strength and hold him well within himself, but that I might not overtake the bison too quickly
and head him back on the difficulty we had just passed—
for I had learned that that was a dodge with which he was
perfectly acquainted. Oh! it was *so* lovely, that wild
ride over the plains, the mighty game in view! Powerful as the sun on that day was, and great as my exertion
from the fractiousness of my horse had been, both Taymouth and myself were as hot as man and horse could
well be (he was white with foam from head to foot); but we
were both in high spirits. It soon became evident to me
that my horse was as free and able as ever, and it was
not long before the bull discovered that I was drawing
on him fast; for, on the thunder of my hoofs reaching
his ears, as I now came furiously up, to my surprise, he
slackened his pace and assumed a more collected canter;
then, when I was within sixty yards of him, he broke
into a trot; a little nearer, the trot became very slow.
His huge head, as well as his tail, was raised; he drew
himself up to his full height, and facing suddenly with a
sort of side swing right about, he offered his shaggy
head to me as I came on, and at once stood stock-still
prepared for battle. As he faced about Taymouth, staring at him, also made a full stop, and for a moment the
horse and his rider and the bull stirred not another inch.

[Spottiswoode & Co.

Printed by]

THE BISON'S CHARGE.

I was then about fifty yards from the splendid game, when on my pressing Taymouth to go on again he at once refused, and as he turned and shied away a little, I saw that the bison with his tail gave that well-understood sign, by raising it higher, that he was bent on charging. Taymouth had then turned his quarters to the bull, and commenced the trick he had of jibbing and rearing when his will was thwarted; so exclaiming to myself "that that would never do, or in an instant we should receive such pressure from behind that the fate of the battle would very soon be decided the wrong way," instead of trying to turn him towards the bull, I dropped my hand, and giving him the spur sent him off a few strides in full retreat; then with the near rein bringing him round to the left obliquely at the bull, I again approached to closer quarters. The bison was aware of the act in an instant, for on nearing him he made the most beautiful dash at me I ever saw or perhaps shall ever see again, flinging the rent turf of the plains far behind him with his short and immensely powerful hind legs; but on the snaffle, faster still away flew Taymouth, when the bull, finding he could not catch me, after a charge of about thirty yards, again stood still.

I was instantly aware of his halt, for in a few strides my willing horse put me at leisure enough to lay my hand on the croup of the saddle and look back, when the instant the bull halted I came round him again, keeping him on my left side; but on nearing him he charged most viciously with the same result as before, and then drawing himself up again stood sullenly still. I had on other occasions, when among a herd of retreating bisons, tried the effect on them of the human voice, and seen the sound for a moment or two scatter them in all directions, so I

gave my foe a view halloa, but he only violently shook
his head as if he hated the sound, though it no longer
scared him, and by the way he received the shout I saw
that he did not intend to retreat any more.   This fact
ascertained, I walked my horse round him; if I came too
near he charged, but at last he contented himself with
simply turning as I turned, to keep his head to me, and
now and then, if he thought me too near, raising his tail.
I looked at the little carbine in my hand, and then at the
magnificent spectacle of the huge animal at bay, un-
scathed and savage and robust in the wildest beauty, and
in my heart I could have been contented with the splendid picture before me, and have permitted the warrior to
have gone back to his sweet-breathed kine upon the plains.
There we were, the bull, the horse, and myself, the sole
living things in sight within an area of many miles, the
only noise the breathing of my horse, or his occasional
snort, which now had no terrors for the foe, and the only
little difficulty that remained was to induce Taymouth to
let me take a steady aim.

Once again, more to inflame my wish to kill and to
stifle any gentler thought than for any other purpose, I
neared the bull, and once more he made a desperate
charge, and once more I retreated, and then plainly read
that the mighty monarch of the desert in his mind had
set me down as a vanquished foe, for he took less notice
of me, and, with a less ferocious expression of countenance, gazed wistfully at the distant hills in which he had
severed from his herd. "The time is come," I whispered
to the silent air; "I must kill, or in these realms my
power to have done so will be doubted." So I raised my
carbine, when Taymouth sprang like a deer for yards on

one side, and continued to toss his head. "Well, then; soho, boy!" and soothing my horse, I walked him in circles round the bull, who now did not even condescend always to keep his horns to me, perpetually raising my carbine as if to fire, till Taymouth got quite used to it, and shied not at all. "Soho, boy," still nearing the bull, when, gently checking my horse to a standstill so near the foe that he again threatened to charge, and was about to turn on me, I shot him just behind and close above the elbow, my horse standing to perfection. He never moved; it was the death-blow; but in an instant again the carbine was reloaded and levelled, and again the conical ball this time gave the *coup de grace*. The monster swayed for a second to and fro, and as he fell dead on his side upon the plain the English death halloa rang aloft and reached the only ear on the plains that understood it.

No sooner had he fallen than my sensible steed absolutely wished to go up to him, when, having permitted him to smell the carcass, I dismounted, and, with the rein across my elbow, and with a steed now thus perfectly made, I sat on the body of the vanquished foe, to pat and make much of my horse, to contemplate the picture, and to scan the scene, in order to ascertain where I was. There was the line of the great creek which in the chase I had crossed, and somewhere on which and up stream I knew my camp to be situated, so my doubt as to being able eventually to reach it was at an end; then also on a distant hill I thought I saw the track on which the waggons had travelled in reaching the creek on the first evening, and all that remained to be wished for was the men and the waggon we had taken out with us to come and secure the robe and the meat; but no living thing was to be seen.

At last two specks appeared on the hills on the other side of the creek; they were horsemen, coming towards me. George Bromfield, in endeavouring to ascertain whither the chase had led me, had heard the "who-whoop" that he had so often heard over Druid's hunted deer, borne on the favouring air, when, directed by the sound, on attaining the brow of a hill, he had made me out standing by a black mass upon the plain. He then went back in search of Major Martin, and together as witnesses of the beginning and the end of the 40 minutes' chase, they came down to the creek, whither I met them, to say where they could cross, and then we returned together to the slain.

We had not been long so situated when Bayard also came down the distant undulations of the plains; with Pape's rifle he, too, had killed his bison at a brilliant shot, whose tongue he bore at his saddle, and then we kept a look-out in case the waggon, on its return to camp, should come in sight. It did so, and with heart-felt satisfaction I gave orders for the robe, head, and all to be taken, as well as the meat. The public have since had an opportunity of inspecting this bison in the window of the Field office, wherein Mr Ward, the naturalist, of Vere-street, has placed him to my entire satisfaction, and my brother sportsmen will be able to judge of the game the plains afford.

Mr Ward has succeeded admirably in preserving the bison in an attitude as sketched by me, giving the exact position and look of the animal, when turning to keep me in his front, his eye fixed on the horse, for a contemplated charge. When the bull in full vigour thus draws himself up, the hump of course looks much less than it does when the crest and head are lowered.

CHAPTER XVIII.

HAVING left the men to take the robe, head, and meat
of the bison mentioned in the last chapter, we went to
inspect the beaver dam in the creek, and very curious it
was. The situation of the holt where the colony lived
was much the same as those used by otters. The entrance was under water, and there were air-chimneys to
the surface of the bank above, for the purposes of ventilation and respiration. There were also places in the
grass where they rolled themselves dry, and a very considerable playground and many paths much used; the
paths tending to cut off an angle of the river, and termin-
ating in a "slide," or widely-beaten place in the grass,

where the animals take the water.   The grass at the
" slide" seemed to be laid all one way, as if the beavers
came with a rush to their favourite element.   Along the
side of the stream were trees, some of them as thick as
my waist, all more or less ready to fall, these curious
animals working at them on the contrary side from
the direction they wished the tree to take, and not the
mark of a tooth in the wrong place.   The dam to pen
back the water, and which made a waterfall, was from
four to five feet high, and as complete and beautifully
erected with weeds, twigs, mud, and grass as anything
could by possibility be.   Had we had time we might
have dug down to and stormed the holt of these beavers,
and had some capital sport with Druid and Bar; but as
the leave of my companions was up, and they were
forced to return to Fort Riley, with a sigh I bade adieu
to the amphibious game, and we proceeded to our dinner
at the camp.

On thinking over my chase of the splendid bull, as
narrated in the last chapter, in all his full power of life
and limb, and bringing him to an unwounded bay, I
made up my mind, that supposing the hunter could
afford to take so much out of his horse, it was infinitely
the most graceful as well as enjoyable way of killing the
great bison.   It had been told me in St Louis, and other
places, that *a corn-reared horse, when on prairie grass, and
nothing else*, could not run down a buffalo, and that a
horse so fed could *not very long hold his own*, even at the
side of the retreating herd;   that no one had ever so
rode down an unwounded bison on a horse so situated,
and that an attempt to do it would be vain, for that a
buffalo, sound and unwounded, could continue his gallop
for many hours.   I am now convinced, however, that

one of the master bulls will not be driven an inch further than he likes, and that if you persecute him too much, and have a horse able to do it, that he will turn to fight long before he is under any muscular fatigue.   To see a bull of this sort halt on his native plain, and dare the close approach of man, under the peculiarly ferocious aspect as well as power assigned him by nature, is absolutely a poem in itself, well worthy the contemplation of any lover of nature or scion of the chase, and I would not have missed its advent for a great deal.   The solitude of the scene, the size and appearance of the game, the immense area of desert around you, and the fact that you are there dependent on your horsemanship and your aim with the *diminutive-looking* weapon in your hand for your life as well as for the successful issue of the chase, altogether make the situation so beautiful that words fail to convey any idea of the pleasurable excitement of the hunter's position.

In one of Mr Bayard's runs at buffalo, when mounted on a steady and perfect horse, the following incident befell him, which will prove how quickly and unexpectedly a buffalo *can* charge out of the direct line in which, at a gallop, he is going, when he is so inclined.   Mr Bayard had ridden up to a bull in a herd, and wounded him, but not apparently with any severity, when as he was still at three-parts speed by the side of the herd, and about to deliver the last barrel of his revolver at the stricken bison, the bull turned to charge so swiftly and suddenly, which the shortness of his muscular limbs enabled him to do, that the well-broken steed which Bayard was riding, perfectly awake to the danger, to avoid the shock was obliged to turn so shortly that he lost his legs, slipped up and fell heavily on his side.   Mr Bayard was

too good a horseman, however, to let go the rein, and
horse and man gained their feet together as the bull, in
his rush, was on them. With that cool determination so
graceful as well as useful in the soldier, gentleman, and
hunter, Bayard resolved to let his horse save himself,
and to bear the risk alone, aided by the one barrel left
of his heavy revolver—a poor chance against the head of
an enraged buffalo—so he loosed the rein. His steed,
too happy to save himself by flight, bounded off, leaving
his master standing erect and motionless, and confronting his hirsute foe. For an instant the huge bull and
Bayard stood stock still on the plain within a few yards
of each other, Bayard with his eye, hand, and revolver
on the beast, but the beast stedfastly staring after the
retreating steed. It was a frightful pause, and Bayard
never stirred hand nor foot, when to his intense satisfaction, the bull (taking no notice of him whatever, his
entire attention having been fixed on the horse) turned
round and trotted off to rejoin the herd, Bayard taking
very good care not to disturb him in that gregarious
intention by any further molestation.

The single-handed chase of mine with the bison gave
me a very good insight into the animal's capability and
inclination, and proved to me that which I had observed
before—that with the size of animals, from the mouse to
the elephant, with very few exceptions, power of observation is very apt to increase with size, and the larger the
brain the greater the intellectual power. The elephant is
decidedly the most sensible beast in existence; and nothing
could be more curious than the fact of the bison couching
in the reeds, with the evident intention of getting me and
my horse to such close quarters as would give him the advantage. The slackening of his speed, too, whenever I

approached him, and his going out of his line to put
creeks and difficulties between pursuer and pursued,
whenever he found himself out-paced, showed that he was
thinking all the time of the work in hand, and the best
means for his own escape. The moment he had thus
placed difficulties between me and him, he invariably put
on all the speed he could, from which speed he again
slackened when he found the difficulty of no avail. The
expression of the animal when at last he discovered that
flight and difficulty no longer availed him, and that he
must fight, was most savagely grand ; and the change in
his aspect, after driving me back several times, when he
deemed that he had succeeded in scaring me from immediate contact, the only tug of war natural to him and which
he dreamed of, was remarkable in the extreme. I shall never
forget the placidity of his eyes nor the expression of the
no longer furious face, when, treating me with contempt,
the bison raised his head, and fixed his longing and
inquiring look on the far-off undulations of the plains
whence he had been driven, and where he knew he had
left the things he loved ; and, I repeat, that had I considered my own feelings, and not what some people would
have been too happy to have said of the event of this chase,
I should have been inclined to have let the noble bull have
taken his chance against other hunters, and have spared
his life ; for I loved the animal kingdom too much to be
needlessly prodigal of life. The event, however, terminated as I have told the reader, and the game was slain.

It was necessary for me some time ago to refer to this
peculiarly successful chase, because the "Jefferson Bricks"
of Charles Dickens had been snarling at me in portions of
the press in the United States, from my first advent to their
land up to the time of my departure, and these "means"

by which, according to their own account, or according to
the account of "Boz," "by which the bubbling passions of
their country find a vent," had described me as having
had what they vulgarly called the "buck-fever," and
" failed with the larger game." No sooner did I refer to
this chase, done in the presence of witnesses, as a very
palpable means to refute their aspersions, than by the same
" means or vent for the escape of the bubbling passions of
their country" out rushed no end of bubbles in the shape
of riled men, all claiming long ago to have done the same
thing, and not only fairly ridden down an unwounded
bison, but to have slain him with a pistol. Had I said
that I had ended the chase with death by a pistol, I have
no doubt they would have claimed the bow and arrow ;
and had I described that as my weapon, they would have
asserted a popgun. Men—and men who knew what they
were saying—told me that thus to run down a bull was
impossible ; I have shown that it was not so, and I have
quite made up my mind that, if a hunter can afford to
take so much out of a good horse, and quality not the
quantity of sport is to be considered, the most sporting
and graceful way to kill a huge and powerful bull in his
prime—I don't include worn-out animals—is to ride him
to bay and then to kill him.

I had hunted over the grounds of the Kaws, who are
not a wild tribe, and over those of the Kiowas, the Chey-
ennes, and the Arapahoos, who are wild—at least so my
friends informed me; and now, alas ! the hour was
approaching when from Fort Riley the heads of my mules
were to be turned towards home, and I should have to
select the best line of prairies to afford me sport with deer,
wolf, turkey, and wild-fowl, till I reached the "settled-up
country." To go further into the Far West was not

advisable for several reasons, the most cogent of which was
money. I had already far exceeded the contemplated
cost of the expedition, and I desired as soon as possible—
the purpose of my sporting mission having been achieved
—to regain a mart for the sale of my equipment, with all
my fine horses and mules, in safe and good condition.
From the experience I had had of my hired men, it was
perfectly evident that, in presence of an enemy, they could
not have been trusted; neither their habits nor their arms
rendered them efficient for any vigilant or defensive purpose; so, taking all things into consideration, as well as
the lateness of the season, I resolved, their leave being up,
to accompany my friends Major Martin and Mr Bayard
back to Fort Riley, and from Fort Riley to take a new
line across the plains and reach the frontier town of St
Joseph, and there discharge my men and attempt to dispose of my animals and waggons.

In one of my previous chapters relating to a day on
which Mr Canterall had wounded a calf, that ingenious
individual came to me after the first run at the bisons, if
I remember rightly, and asked me to permit George
Bromfield to lend him his revolver. It was one lent to
me by my friend Capt. Bathurst, of the Grenadier Guards,
and which, for that reason, I valued very much. Permission being granted, as I chanced to be in a very good
humour, he stuck the weapon in his belt, and then we
hunted again. At the end of the sport Mr Canterall
informed me that he was very sorry to say that in the chase
he had lost my pistol. Had I known as much of him
then as I know now, revolver in hand I would have
searched every pocket he had; but, at the moment, I
passed the fact by, and am perfectly certain that I made
an immense mistake in believing a word he said.

On the morning of the 14th of October orders for the countermarch were issued. We breakfasted at daybreak, and thus ruled the line of duty for the day ; Major Martin having expressed a desire for another Buffalo robe, Bayard determined to take two of his men on mules, and to proceed to where he had left the carcase of his bull killed on the previous day. Major Martin, with my ambulance, myself, and one of the Fort Riley waggons, were to await at our camping-place the return of Bayard, and the other waggon in company with my baggage-waggon, with Mr Canterall on my pony, were to proceed on the line of march, and to halt at the creek at which we had encamped as we came out. Bayard was confident that he should be able to rejoin us, robe and all, by nine o'clock ; but, as must often happen on that boundless extent of markless land, he was a very long time in finding the spot he sought, and when he found his bison, and the men had skinned it, on putting the raw hide on the mule, the mule was so terrified that he broke away and fled over the desert. Some time elapsed before the mule could be recaptured, and the day became so far advanced that we were suspicious of interruption by Indians, and I had it in contemplation to go out in search of Mr Bayard's portion of our expedition.

While we were waiting the return of Bayard, I took Brutus and my double shot gun to look for game in the vicinity of the camping ground, but met with none, a prairie hawk being the only bird I killed. On returning Major Martin informed me that from the line of march taken by the waggons and Mr Canterall, he had distinctly counted twenty shots, and that he had not the slightest doubt but that fellow Mr Canterall had sighted

buffalo, and was, in direct contravention of his orders,
disturbing the ground. I had not long sat on a gently
rising eminence, talking to Major Martin and Willie,
when, from the hills over which our waggons had for
some time disappeared, we saw a single buffalo descend
in our direction in such guise as proved him to have
been disturbed. Having caught up my rifle, on seeing
that he held his way for the creek above us, I ran to
meet him, and having attained a spot that seemed to be
in his line, I lay down beside the bank of the creek
itself, and on a bison path. The bull for a time kept on
towards me, when, to my indignation, just over the
brow of the hill behind him, I saw that unmitigated
rascal Mr Canterall on my poor little pony. It seemed
that, as soon as I saw him, he feared, too, that he should
be seen, for had the bull turned at him he could not
have pulled up shorter, nor rode away faster, than he
did when he found himself unexpectedly in sight of the
camping ground, and very likely within long reach of
my rifle. The bull also turned away, and I went back
to Major Martin and told him what I had seen. Feeling
uneasy at Bayard's prolonged delay, mounting Sylph,
and taking my carbine, I rode a considerable distance to
a rising ground which had command not only of the
camp but over the vista of plains, in the undulating
grounds of which I knew Mr Bayard to be, so that I
should gain any tidings of his approach, and be able too
to signal to the camp if I saw anything to call for
further aid. After waiting some time, I was delighted
to make out the right number of specks coming towards me, which answered to Bayard and his two
men; and by noon Mr Bayard had rejoined us with

the robe, and we commenced our march after the other waggons.

As we expected from what Major Martin had heard and I had seen, the whole line of march had been disturbed by Mr Canterall, and we saw no game of any kind that we had a chance of having time to approach, one large herd of buffaloes so far off that we could just distinguish what they were was all that came within our observation. We then overtook the other waggons, and reached, just at dark, our camping-ground at Chapman's Creek; and when I asked Mr Canterall what he had been doing, he swore he had only shot at buffalo close to the line of march, and that he had killed two, but as he could show me neither tongue nor tail, of course it was a lie. The men and Mr Canterall had killed a buffalo on the line of march, and that was all. I then told the fellow that the pony was evidently tired to death, that I had seen him riding after a bull, and that it was totally impossible to believe a word he said.

At dinner Bayard informed me that he intended to ride on to the fort next morning, about two o'clock, a.m., so as to precede our arrival by many hours; his excuse was that his time of leave was up, but I am now sure that the kind motive which took him away so early was to provide an entertainment for me. I had received an invitation to rest as long as I liked, on my return to the fort, at the quarters of Major and Mrs Martin, and also at my former kind entertainers, those of Major and Mrs Wassells, when, on finding that Major Wassells would be absent on a court-martial, I accepted Major Martin's proposal, supposing, too, that from Major Wassell's absence it might be inconvenient to his lady to entertain me

20

Bayard having started to ride the distance over the solitary plains at 2 a.m., at daybreak we struck camp and followed in the same direction, when in the course of the day the following amusing incident occurred. Not apprehending the presence of any game, old Druid was at liberty, and as usual in his favourite place, trotting a long way ahead of the ambulance in which I was; suddenly out of the grass, but on the wrong side for him to catch the wind, a very large wolf arose and stared at the hound. I soon possessed myself of my Manton and heavy deer cartridge, but ere my ambulance arrived within shot, while Druid passed without noticing the wolf, the latter cantered off and was lost among the undulations of the plains. Druid continued to trot on ahead, when, alas! too far off to be in reach of the cartridge, on the road and coming to meet him I saw another large wolf, when as Druid reached the summit of a little rise in the track, the wolf came up on the other side, and they met face to face. Bang at the wolf went the old hound, and off the wolf set with Druid racing after him over the plains, and before my mare could be brought up—the Boh-hoy who had her in charge having listlessly loitered in the rear— the wolf had shut out the hound from view, precisely as a fox would have done, by availing himself of hollows and inequalities in the plains, and old Druid was flinging away vigorously on his scent, but, to my astonishment, in perfect silence. I am certain that the hound was not sure whether it was a dog, or what it was that he was on, or if he was right in chasing, never having felt such a scent before, and always having been steady to deer. I have often, when he was short of work, seen him run and open on a fox, but on the wolf he was

silent. On finding that my mare was a long way behind, and that old Druid, in his chase of the wolf, was
far out of sight, I begged Major Martin to follow him
up, when, leaving orders with my ambulance man to
send the mare after me, snatching up my gun and
straight English hunting-horn, I ran off on the line
Druid had taken. Like all animals of his class, the
wolf had gone a down wind course, which luckily left the
recall of my horn available to the ears of Druid; so,
having reached the crown of a hill, which enabled me to
distinguish that Major Martin had pulled up, I touched
my horn to get back the hound. It was not very long
before I caught sight of a dark speck coming to me,
varying in his gallop according as the sound of his well-
known horn might reach him, and Druid returned, having had a very good gallop to put him in wind, for
expected deer.

After this little diversion, I bagged a brace of grouse,
and trotting my ambulance freely, we eventually reached
the Kansas river, on which the fort stands, soon after two
o'clock p.m., when, having requested the person living
by the bridge to direct Mr Canterall to a good place for
my encampment on the same side the river as the Fort, I
repaired to the barracks and found that Bayard had
safely arrived, and awaiting me the heartiest and kindest
reception. Why Bayard had trotted on was very soon
made evident, for in his quarters he entertained all his
brother officers and myself to as good a dinner as any
man could desire to sit down to—lots of champagne and
excellent roast beef, thoroughly seasoned by that sort of
good fellowship which should exist between soldiers and
gentlemen—and a more agreeable dinner of the sort I
never desire to share.

We were all intensely amused by my dear old dog Brutus, who had accompanied me to the Fort. So many were the joints and so hospitable the cheer, that the very excellent and handy black servant who waited had not convenience sufficient whereon to place all the good things, so, on a chair behind me for a time he put a hitherto uncut and immensely hot and excellent ham. Brutus, who is beautifully behaved, was standing by my chair in the anxious hope that he should be told to fetch something. The telling he desired, however, not coming, he thought he would at least make himself of some service; so he walked up to the ham with the full intention of bringing it to me. A heavy bump on the floor made us all look round, and then we saw the dear thing licking his lips over the savoury viand in an endeavour to cool them, the ham which he had directed a few steps towards me having been too hot for him to hold any longer. With the funniest face looking towards me, he seemed to explain what had been his intention, and to regret that he had found it too hot to be accomplished.

Sunday, the 16th, proved to be a drenching wet day. At church, a mere room in the barracks set apart for that purpose, my reverend countryman from Bristol, who officiated, gave us an excellent discourse; but, to my surprise, the congregation did not consist of above ten or eleven persons, and those were chiefly officers and their ladies. Of course this created some notice on my part, and I learned that the soldiers generally, being Irish, were Roman Catholics; and that was the excuse or otherwise for their non-attendance at church. Well, then, if the United States in their boasted desire to maintain freedom (I suppose in the boast they include freedom of conscience,

as well as a general license to be democratically disobedi-
ent to all that is lawful and just) find it in their Government good to enlist a foreign army, surely they ought to
assign from the State funds sufficient to supply that army
with a place of divine worship, and to maintain a priest.
As I have before remarked, it is the custom in the United
States to go a-head—by what means that is done is imma-
terial. As their own natural-born sons will not serve as privates in the ranks of the army, but as an army they must
have, they get it by foreign enlistment; but whether the
component parts of it go to the devil or not, after their five
years of enlistment are over, is not worth the consideration
of the representatives in Congress brought together. It
is a fact, after you leave the frontier towns, men, whiskey,
and tobacco go a-head—roads and religion are left
behind.

On the following morning, Monday, the 17th of October,
we had settled to take Druid, and draw "Smokyhill
Creek" for a deer, but it again set in to blow and rain,
so that we agreed to postpone the attempt till Tuesday.
The weather had become bitterly cold, there were large
fires kept in all the officers' rooms, and I began to ask
myself where was "the Indian summer?" On the 18th
of October the weather cleared up, and it was very fine,
so we set out from the Fort to get Druid, when, on reaching
my camp, I found that Mr Canterall and Tom had taken
two mules, and on the preceding day gone out, against
rules, on their own hook, to see what they could kill;
they killed nothing, but they started a buck, so at all
events I became aware that there had been a deer in the
vicinity.

Having obtained Druid, and accompanied by some of
the ladies from the barracks, we then began to draw the

tangled forest on the banks of the river, and anything half
so severe for horse or hound I never saw. The growth of
wild vines and a sort of briar covered with thorns, were
strong as ropes, and would hold a horse, while the up-
rooted trees, cast in all directions, and grown over by
creepers, were imperviable to a hound, and impossible in
regard to any fast pace. The banks of the wide river,
too, for a vast extent, as well as in many places, were so
steep and high, that if a hound went down to drink it
was impossible, for hundreds of yards, for him to get
out again.

Having drawn a considerable distance, and seen the slot
of a deer, Druid at last became curious, and ran his nose
along the twigs in his path that might have brushed the
body of a deer, and I became fully aware that he knew
there was one not very far off. This I told to Bayard,
who was the only one, with Master Willie and George
Bromfield on my pony, who kept me in company. At
last Druid began to " feather" in his mysterious way on
traces of deer, when, by the tangled cover, we were for a
time separated. Presently Bayard called out that Druid
was still more fresh on the line of something, and that,
though he had encouraged and called to him to come out,
he would not leave the bushes where he was. I had
scarcely called out "all right" when the forest rang with
such a roar from the faithful hound as made the woodpeckers fly again; and ere he had roared a second time,
and while I was cheering him to the echo—for I well knew
he had roused a deer—George gave a view halloa. Hound
and deer came on to within forty yards of me, and I stood
stock still with my gun prepared; in another instant I
could hear the rustle in the bushes made by the deer, when,

just as I caught a scarcely sure glimpse of her—again, O ye fiends of the Hartz mountains!—a dell in the ground enabled the doe to plunge out of sight and I could not fire. Hound and deer then took a narrow ring, when, strange to say, precisely the same thing happened to Bayard; the deer came within ten yards of him, and again baulked the shot by suddenly bounding down a hollow in the cover. The scent was very bad, as I suspect it always is there in dry weather, from the immense accumulation of withered leaves and the quantity of dust driven in by the winds, both burnt from the prairie fires, and natural, from the lightness of the surface soil itself. What became of that doe, or how Druid lost her, I am unable to say; it was just as if she had sunk into the ground, and no cast that I could make or the hound in his sagacity imagine recovered her line again.

We then drew over a considerable line of woodland without a second find, though there were stale traces of deer, and returned disappointed to the Fort. The time then came when I was to bid adieu to so much that was graceful and high-toned; and, full of regret at parting with such agreeable friends, I gave order to Mr Canterall to prepare for a start towards St Joseph, instead of returning by Kansas city, which order seemed to fill the recipient of it with the deepest dissatisfaction. During the wet days, since our return to the Fort, when we could not hunt, I had had Mr Canterall up with two of his men to work, under my immediate supervision, at the buffalo robes, which he had miserably neglected, and every hour made me more dissatisfied with his conduct. All being then packed up, with a prairie dog or marmot added to my live stock, I took leave of my most kind friends, and with a promise to keep up my communication with them by letter, and a hope that I should some day see them in England, I went to bed, intending to start on the following morning as early as possible.

CHAPTER XIX.

IT was on Wednesday, the 19th of October, that early in the morning I bade adieu to my kind friends at Fort Riley, and proceeded on my route to St Joseph. On the march some portions of the scenery, particularly on the banks of the Cat River, were prettily wooded, but the country was destitute of game, and the only shot the day afforded was at a common teal, which was added to my larder. We camped just before dark at Big Blue River, near Manhatton. On the road as we travelled we picked up a cream-coloured mole, such as we have in England; in America that colour, as belonging to these animals, is, I believe, more common than with us.

On Thursday, the 20th, as there were a good many

creeks on the line of march, I resolved to draw for a
deer, leaving the waggons to continue on their direct
course to the camping-place, and to join them there by
night. A more likely place for deer than the covert we
tried on the banks of the Big Blue River, I never saw, and
in that covert there were several of the common English
rabbit; but though they were close beneath the feet of
my brown mare. Sylph, whom I rode because Taymouth, being considerably "tucked up" after his chase
with the bison, was not fit to go, I spared their lives,
inasmuch as the deer cartridge with which my Manton
gun was loaded would have blown them all to pieces.
As Mr Canterall, George Bromfield, and myself were
riding about through the covert, three wild geese passed
over the river very nearly in shot, and gave me a hint
that the time was approaching when I might expect to
meet with flocks of these birds in all appropriate
places.

Having tried a vast deal of covert, but all in vain, I
came to a cabin where a young man was leaning over
a gate. He accosted me civilly, asking the question if I
was "out on a hunt?" when having replied in the
affirmative, I asked him if there were any deer or wild
turkeys in the neighbourhood, when he assured me that
they were so scarce that it was almost hopeless to look
for one of either kind. This off-hand assurance, given
in perfect sincerity, was hardly out of his mouth, when
an older man, who had partly heard my question, and
perhaps the reply, came hastily out of the cabin, and at
once redeemed the character for romance that pervades
certain classes of that "pattern republic," by reiterating
my previous question: "Any deer or turkeys? Guess
there are! Turkeys? I reckon 'em by thousands here-

abouts; and as to them deer, they're fit to run clear over me every time I rides out. Most afeared my hoss 'll come to grief, they're always so nigh a coming bounce agin un. Yas, sir!" "Thanks," I replied; "your news, sir, is certainly cheering; but if I find one deer I shall be content, but that I very much doubt." Thus saying, I passed on, and shortly after met another specimen of the free citizens in the shape of the tallest, raggedest, ruggedest, hairiest man I ever saw, but who spoke the truth, and told me it was next to impossible in those parts to find a turkey or a deer. He, however, indicated the line of creeks where I should be most likely to meet with success. We then came to an extent of prairies, through which wound several lines of narrow and prettily-wooded creeks, when, ordering Mr Canterall to the opposite side from me, we took the one that seemed to promise the best covert. Old Druid drew the covert very well and for many miles without the smallest touch of a deer, when all of a sudden bang went Mr Canterall's gun, though that respectable individual backed it not in any way with his voice, so I called to him to know what he shot at, and he replied, "A wolf here in the long grass—I think I must have hit him." "Jump off the pony, George," I cried, "get to the spot and holloa on Druid and Par." Old Druid was close to me, staring in the direction of the report of the gun, and listening to know if his services were required, and when George had attained the spot indicated by Mr Canterall, and given a holloa, with a yell of anxiety the hound plunged down the steep bank, and in an instant I heard George giving him the usual gentle encouragement to pick up a line of scent. "Well, can't you make it out?" I said, after a long silence. "No, sir," replied

George, " the hounds won't have it; there is no scent of any sort here, and if there had been anything on foot it must have crossed the bend of the creek, but there is nothing to be seen on mud or sand, and it's all nonsense." How Mr Canterall relished this very plain counter-assertion to his cry of " wolf" I know not; he said nothing, and I suspect he let his gun off by accident; we then continued on our draw.

Some time after this I perceived at last, by the peculiar wave of his stern, that Druid had found the very stale usings of a deer, when a little further on, on observing the stern still more busily at work, I rode to the spot, and then found Druid arrived at two lairs, probably of one and the same deer, who had shifted his position. From this spot we continued again for miles without a " touch " of any sort, and it was getting late in the day when we left the creek we had been trying, and were proceeding in a more direct line for the camping-place on each side of a low dry ravine or small creek, or rather winter watercourse, in which there was now nothing but rank grass. I was obliged to make a little detour from the line of this cover to avoid a steep rocky promontory, around which I had just arrived in my way again to ride by the side of the long grass, Druid and Bar both at my heels, when right in front of Mr Canterall's horse and within thirty yards of him I saw a young doe spring from her lair, and run straight away from him. " Bang ! " went Mr Canterall, missing the fairest shot I ever saw, and then failing to make any use of his second barrel. " Holloa on the hound," I shouted, but, alas ! the ignoramus knew not how to do it; so with all the haste I could I reached the spot, and away with a roar went Druid, but still faster and perfectly mute went Bar over a wide stretch of beau-

tiful prairie, beyond which, after a brisk gallop, I could
see we must reach another wooded creek. Sylph needed
but a shake of the rein to fly after them, and a prettier burst
was never seen. After surmounting a gradual rise in the
plains there was a very extensive slope in the land, so that
I could command a view of all that took place. Bar was,
at least, a mile ahead of Druid, and having come to a
check, he was casting to regain the line, his head down,
and questing in some longish grass. I guessed the doe
was down, and made all haste to get up within shot,
but before I could do so I saw her rise just at the moment,
when the only bush that grew near the spot intervened
between her and the eyes of Bar, or there would have
been an opportunity for the loveliest course ever seen,
and she gained the start of a minute or two before Bar
recovered the scent and again began to run. The check
let in Druid, who never hesitated, but started again
almost side by side with Bar ; the latter, however, again
drew considerably ahead, and held the lead for some
distance. Gradually, then, as the speed of the bloodhound held its own, Bar's began to fail, and the space
between them visibly diminished. In a short time they
were neck and neck, straining every nerve to my cheer,
as Sylph flew by their side, and then Druid took the lead,
which he increased at every stride, and this lasted till Bar
stood still, and George picked him up, while I kept
merrily on with the bloodhound. We had now had a
very good run of three quarters of an hour or thereabouts, when the chase turned from the open plains short
for another creek, and guessing that the deer would wait
for us I hastened up, but, alas ! the demon of ill-luck was
resolved that day to thwart me, and I had to lie out of
my course to avoid a considerable swamp. By Druid's

tongue I knew that he was in water, and as he occasionally checked I was convinced that the deer was trying to
baffle him by "taking soil," and going in and out of the
water, and I would have given worlds to have arrived
on the immediate spot. Before I could attain it, however,
and at the exact place which I was endeavouring to reach,
the doe made her appearance behind the hound, when at
considerably more than a hundred yards, and when going
at speed, I took a random shot, but of course in vain, and
she disappeared again into the creek, when, if Mr Canterall had had his wits about him, and if he could have held
his gun straight, he might very possibly have been of
service. He ought to have got on considerably ahead, so
much so as to have made himself quite certain that he
would have kept the doe between him and me; instead
of this he made for the exact spot in which the doe had
disappeared, and as in such a situation he was infinitely
more likely to have done harm than good, I cried to him
to stand still, and then I called on Druid. Unfortunately
at this moment two fresh obstacles to haste arose: one
was that the banks of the creek, in many places from the
water's edge, were so rocky and steep, that having hunted
the deer into the water, if Druid missed some narrow land-
ing-place by which to ascend, he could not find another for
a long distance; and the other was that we had crossed the
line of a fresh deer. After considerable delay, however,
I regained the obedient attention of the hound, and laid
him on the hunted deer, and by degrees, but more coldly,
we hunted up to George, who had viewed the doe going
back over the plains in the direction whence she had
come. The scent, however, with the decline of day, got
much worse, when as we were still an unknown distance
from the camping-place, and the shades of evening were

coming on, and as I saw that Mr Canterall knew nothing
of the locality, I ordered George to take up Druid, and
we resigned the chase. The sides of the plains were
here considerably broken with winter water-courses and
some springs. We saw some prairie grouse, and I ought
to have had a shot at a flock of English teal, which rose
suddenly out of a small hole of water—so suddenly and
with such a flutter that their flapping of wings, coupled
with the presentation of my gun, caused Sylph to shy
so much while she was on dangerous ground, that in
steadying her I could not pull a trigger with any hope
of success.

> Dark and more dreary lowered the gloomy day,
> Wild were the plains and doubtful grew the way.

So doubtful did the ground in fact appear that I gave
Mr Canterall a caution, and warned him that I was convinced some dark green grass we were approaching was
of a treacherous character, and that before he crossed the
swampy-looking place where this greener grass grew he
had better dismount and sound it with his foot. Not
attending to this suggestion of mine, the fool at once
showed that he had just that amount of courage which,
though it sufficed to get him into a difficulty, had not
stamina to bring him successfully out again; a line of
conduct he would surely have pursued had we met hostile Indians; so to show me this dangerous amount of
dubious pluck, with a careless exclamation of all right,
on he went, and in an instant my bay horse Kansas
sank with a floundering struggle in the swamp up to
his tail, and Mr Canterall, tumbling off in abject terror,
let go the bridle, and never ceased rolling out of an
apprehended danger's way for several yards. The
consequence of this unhorsemanlike act was, that

while Mr Canterall was performing these cowardly
evolutions, the horse righted himself, got out of the bog,
and went trotting off over an endless extent of recently
burned prairie, in a twilight which rendered it very
doubtful if I should ever see him again.   Casting my eyes
upon the bog, while I commanded Mr Canterall to stand
still if he wished to continue rubbing his back, and not
attemptto run after the horse, I saw a drier-coloured grass,
and within thirty yards of the spot where this mischief happened, Sylph carried me over without a struggle. I then
quietly followed the bay horse, who, when he saw a boundless space of prairie in front of him, all dry and burnt as
black as a coal, came back to my mare, when, seizing the
rein, I once more delivered him to Mr Canterall.   It was
dark when we got to the camp at Rock Creek, hounds
men, and horses very much tired, when I enjoyed both
my dinner and my blankets, though the waggon was not
half so comfortable as the tent lent me by my friends
while at Fort Riley.

On the 21st of October we had breakfasted and struck
camp by eight o'clock in the morning, and travelling
through prairies occasionally settled, we reached the
French Indian settlements about one o'clock, and subsequently " Pottowadomy," so pronounced. The population of this considerable village was of all shades of
colour, from white to red, and all of them educated in
the Roman Catholic faith. On the way we passed
through miles and miles of burnt prairies, and indeed
before I left Fort Riley the horizon for an immense
extent had been one vast area of ruddy reflection. More
than once on the road we had had to pass fires still burning, the beaten track enabling us just to pass; the
sparks, far from agreeable, were flying all round my

ambulance with its little magazine of gunpowder. In journeying on this day to the Indian settlements, as I was seated in my waggon, my shot-gun loaded with the common charge for feathered game as usual across my knees, I saw in a little garden attached to a cabin close to the track-way two great birds sitting perfectly motionless, and of a bluish colour, shaped like the common heron, but much larger. I got out of my waggon, and exclaimed to my men that they must be tame birds, and therefore I threw my gun into the hollow of my arm, and was walking on, when up rose the birds, and my men shouted, "Not tame—they are wild!" With the first barrel of Pape's shot-gun I killed one, but at the distance at which the other bird was, though the small shot struck him, it failed in effect, so we only bagged one of these large "blue cranes." The breast of this bird was very fleshy, and when dressed, not bad to eat. The crane and one prairie grouse was all that I bagged on this day.

A bed to sleep in being far preferable to a waggon, and having heard from the officers' ladies at Fort Riley that there was a very nice, clean, and attentive female at Pottowadomy, who kept a lodging-house on the hill, I resolved to send on my baggage waggon and camp (all but my ambulance and dog-cart) to a place called Silver Lake, a few miles beyond the Indian village, and thinking, from the name of the water by which the men were to encamp, there might be both fowl and fish, I desired Mr Canterall to await my arrival at that spot, as, if I saw a prospect of good sport of any sort, I might probably remain there to enjoy it. These matters being arranged, I paid a complimentary visit to the chief priest of the

21

Mission, Father Schultze, and presented him with some
cans of preserved oysters and sardines, asking his blessing, and inviting him to pay me a visit at my lodging,
which invitation he declined on the plea of more serious
engagements, but he sent me some little presents for Mrs
Berkeley, who, I had informed him, professed the same
faith as himself. And that it would be a satisfaction to
me, to take home for her a reminiscence of the
kind. The lady at the lodging on the hill was most
attentive, clean, and obliging, with as great a horror
of the vice of spitting as I had; and what between
the viands she possessed, and those I brought with
me, and her clean and comfortable room and bed, I felt
to some extent in clover. My three mules were put into
a shed, wherein two of my men also slept to take care of
them, as there were no locks; the dog-waggon, with
George and my ambulance, guarded by the dogs, being
drawn up alongside. Old Chance being the least watchful, I permitted him to sleep in my room.

On Saturday, the 22nd of October, I left the French Indian settlement a little before nine in the morning, expecting to find my camp, as I had directed, waiting for me at
Silver Lake; but when I arrived there, where the camp
should have been I saw but a single horseman, who
proved to be no other than Mr Canterall. On asking
him where the camp was, in that whining voice which
he always adopted when he lied, or when he thought I
should be angry, he replied that he had sent it on, having first inquired at yonder cabin if there was any
sport for me or any fowl, and having received a reply
in the negative, he thought I should wish to go on
somewhere else. As may be supposed, from my being
aware of his falsehood, I was very angry, and, pointing

with my hand to some wildfowl on the water, I asked
him why he had not used his own eyes; then standing
upon the seat of the ambulance, I took a view of the
great extent of plain, in the hope that the waggons were
near enough for me to send the scoundrel after them to
bring them back. They were, however, out of sight;
so not wishing to give the mules more work than necessary, on account of keeping them in saleable condition,
I put up with the affront, and resolved for the future to
give my orders to the men. Tom, who was not yet
gone far on, having charge at that time of my horses,
and who really loved sport, for the first time then spoke
out in regard to Mr Canterall's delinquencies, and said
that Mr Canterall had made no inquiries at all, but that
he had hurried on the waggons because all he now wished to do was to get home as soon as possible.

Had Tom not been afraid to speak all the truth, he
could have added the reason of this haste to get the travel over, and that reason would have been that Mr Canterall was in terror of his life, that I should detect his
thefts on my wine and other things, as well as that
perhaps some of the men, who all hated him, but were
afraid of him, might prove that which some of them
were well able to do, that he had cheated me at every
purchase made by him of corn for my horses. Resolving
to intrust this fellow with no more orders further than I
could see them obeyed, I then continued my journey,
overtook the baggage-waggon, and arrived early at Mud
Creek, where we encamped. While dinner was getting
ready a lad came to us to know if we wanted corn, and
while in conversation with him as to the game in that
vicinity, he said there were " loads of coons about the
creek," and that he and others were going to hunt them

that night. On this I asked him if he had any coon-skins, and he replied, "Any number," so I told him to bring me the best he had, and I would buy one or two. The lad went away, but of course, from the independent method in which he had been "raised" as to the truth, he had no skins; but we got some corn and very nice milk and butter.

Taking my shot-gun and Brutus, I then strolled up the creek, and bagged a brace of quails or partridges, and the common English buzzard-hawk. The little stream in the creek was full of chub, as they appeared to me, of from one to two inches long; but further than a few quail and quantities of woodpeckers and hawks, there was nothing else to induce any sportsman, however keen, on an intensely hot afternoon, to wade through the tangled covert.

On the 23rd we left our ground a little before nine in the morning, Mr Canterall uselessly ill again, as well as one of my other men, with ague and fever, and George Bromfield also was not so well. On the march I killed four birds which my men called "snipes." All "waders" with long bills, of whatever description, are so called by the generality of Americans; but the birds really were a species of the redshank, only larger; the legs of the birds were yellow, those of the two old ones darker than the others, which were evidently their offspring; the plumage of all a mottled grey. At three in the afternoon we reached Hickory Point, where we thought of encamping about a mile from a cabin and a corn-field. It being early, before pitching our camp I sent Mr Canterall on to the cabin to inquire where the best grass was, and if we could have any corn. Having obeyed his orders while I looked at him, on leaving the cabin-door he indicated with his hand which way we were to go, when I observed that he was cautiously followed from the cabin by a lean old sow, who, it struck me from her actions, was perfectly well aware of what we were going to do. Having selected a spot for camping on by the side of the brushwood of the creek, as usual I strolled out with Brutus and my shot-gun, but without finding a symptom of any game.

CHAPTER XX.

IT was nearly dark when I sat on the seat of my omnibus to dine off a brace of broiled partridges, sardines,
and fresh butter, and while I was sipping my sherry the
evening closed in calm and hot. With the darkness
also appeared no less than nine different prairie fires,
two of which were not very far from my camp, and to
one of these directly in the eye of the wind I particularly addressed my attention. Thus encamped on the
plains in a dark night these fires have an awful appearance, but they are not half so dangerous as novelists
depict them, or as the vivid minds of some of the
Americans themselves would induce you to believe.
The grass of the open plains cannot hold a fire long if

there is a wind, because the grass is very light and dry, with very little bottom to it, and the wind carries both the flame and ashes away. The only danger is when the fire runs into the heavy grass and brush-wood of a creek; then if a man encamped permits himself to be thus caught, everything he had that could not break loose and run away would be destroyed. Supposing the wind to have driven a considerable fire in my direction, the fire extending for miles, prohibiting any idea of outflanking it, I should immediately have called all hands with lucifer matches, a large quantity of which should ever be *in the commander's possession*, and have fired the plains at my foot, and for a space have followed my own fire. An act such as this, when resolved on in time, puts all danger out of the question.

On the evening of which I speak I sat watching the fires till the breath of wind went entirely down, and then retired to rest. From a very sound sleep I was suddenly aroused by hearing a mule start violently to the end of her larriet, and commence a succession of snorts at some object that offended her. Brutus and Druid too heard the same thing, and flew out angrily to the extent of their chains. Not having any faith in the fidelity of the watch on duty, I arose, pulled on my snake boots, took Pape's breech-loading rifle in my hand, girded my *couteau de chasse* and revolver at my waist, and crept quietly from my ambulance. The moon or the stars were up, though obscured by the smoky sky, and my ear having directed me whence the snort from the mule came, I soon arrived sufficiently near to observe that one of my best mules had retreated to the full extent of her larriet, and in terror still had her eyes fixed on something in the long grass.

Having been told that mules would betray this terror at even the smell of Indians, I brought myself in line with the mule's quarters, in an endeavour to keep her between me and the object she thought she saw, so to conceal my advance, and I thus came up close behind her. I peeped round her loins, but could see no living thing; yet still her long ears were stuck out in one direction, and at intervals she continued to snort, and it became quite obvious to me that the mule saw or smelt something that she was afraid of. Standing thus, with my rifle cocked and ready to my shoulder, my eyes became more reconciled to the darkness, and I detected a very slight motion of some shadowy substance in the grass about the height and just the colour of an Indian upon his knees. The rifle rose to my shoulder, but still I withheld my fire, when, oh, ye gods! the tone of my anticipation was considerably lowered when the long, low, dark thing, almost concealed in the grass, gave a most unmistakeable and swinish grunt, and I discovered that it *was* the old sow I had seen follow Mr Canterall, in the act that made her follow him, that of eating the corn I had bought from her owner to feed my mules and horses. On this I drove her away, and, retreating to the watch-fire, roused the man in charge from his habitual state of lethargy; but the camp was disturbed all night by that sow and a lot of her offspring, who had joined her, and who, from coming too near my ambulance, at last caused Brutus to break his chain, though, when loose, nothing would induce him to give the pig a good shaking by the ear. So annoying was the old sow, that though I turned out all hands to drive her off and flung large logs at her, she returned so often that I had a great mind to shoot her in defence of my corn; but, on reflection,

I thought it being for but one night, I would put up with it.

An immense joke was then had, and, if I recollect rightly, it was with Tom or Wallace, and it was that in order to deter the repetition of the sow's visits, I desired one of my men to put a very small quantity of powder and shot into his gun, just to tickle her. He however put so little powder, that when he fired it had but strength enough to throw forth the shot with no greater force than would have been achieved if flung from a man's hand. The consequence of this abortive attempt was, that the sow absolutely ran back to where the shot fell short of her in the grass, in the expectation that we might have thrown her something to eat. We all passed a very restless night, and, while I lay awake, when rosy-fingered morn first tinged the eastern horizon with a delicate grey, it was curious to hear from distant portions of the prairies the low hollow crow of Cochin China cocks, proving that though I had seen but one cabin, there must have been many more in that apparently uninhabited region.

On Monday, the 24th of October, we left Hickory Point at a quarter past eight, and in and about the first corn-field we came to there was a pack of at least 300 grouse. Observing them to settle among the corn-shocks, I took Chance and Brutus, and Tom also with his gun, and tried in vain to get near them. Having observed that they flew in the direction of our march, and towards some prairies where the grass was very thick, I determined on going their line, ordering my ambulance to proceed some way, and then to wait till I came up to it: and I soon saw that, supposing we had shot half a day, I should completely have outwalked my attendant. As it was, when-ever we came to any drinkable water, however bad, Tom lay down to refresh himself, while I felt as cool as if I had not gone a hundred yards. I obtained only one shot at a single grouse, and then got into my waggon. A little later in the day, while we passed through a wooded creek, a covey of partridges rose and pitched in the brushwood on the side of a hill. On this, with Chance, I went after them, attended by George and Tom with a gun. They had a shot or two, taking turn about at the partridges, but neither of them got a bird. I killed one, and having marked the covey down again, Chance came to where they were, and behaved beautifully. There was very little, or indeed no scent, the cover dusty and dry, and the day intensely hot, and the setter ran into the midst of the covey before he was aware of them, and springing two birds dropped at once on the ground. The partridges that rose, not knowing what Chance was, immediately perched in a low tree over his head, and peered down at him as if he was a vermin. On this, being well within shot, I killed first one, and then the other, dropping them from the boughs right on Chance's head; but he, knowing that there were others in the grass close to him, never stirred. The covey subsequently rose, and I got a brace more, and then rejoined the waggons.

At a little before three in the afternoon we reached a pretty little creek, with a nice stream of water in it, in which were a good many small fish, about two miles from the town of Achinson, on the Missouri river, and there pitched camp for the night. I had not been long arrived at the camping-place when a settler joined me, from whose cabin I obtained corn, and excellent butter, milk, and eggs, and received the information that horse-theft was rife to any amount in that vicinity, and that only a few days before, in broad day, a horse, bridled and saddled, and hitched for a little while to a hook in the wall of the cabin, had been carried off. On hearing this, I issued orders to my men to picket my horses directly in front of my ambulance, and the mules also, in as close proximity as possible, and to keep a good look-out.

It was a very pretty sight, on that beautiful evening, to see Taymouth and Sylph quietly whisking their long tails at the flies, while they ate their corn close to me, with the bay horse and my favourite mule mare very near them, the rest around me, and no other sound to be heard than the tinkling of the pony's gathering bell, and the chorus of frogs, crickets, and grasshoppers of every sort and kind, which made the cover close to me alive with minstrelsy loud as the song of birds. It was very enjoyable, all this; and when I dined, on the seat of the ambulance, on broiled partridges, and a couple of hard eggs, hot rolls, fresh butter, cheese, sardines, and potted cherries, with sherry and brandy, and good water for coffee at command, I looked at the watchful and affectionate faces of my dogs, and as far as the hour went, I thought we had not much reason for complaint. On the following morning I had resolved to trot on ahead with my ambulance, and reaching St Joseph early, there to prepare stables and good accommodation against the arrival of my camp; but a most agreeable intervention of sport prevented my putting that idea into execution.

We were astir before daybreak in the morning, and by half-past eight were in the town of Achinson, waiting to be put across the river. This feat having been duly accomplished, I found that the greater part of the day's journey lay through the primeval forest on the banks of the Missouri, the forest trees very large, and the scenery, as a woodland, wild and beautiful. In wending my way through the woods the track came suddenly on the end of a large sheet or lake of water, when from my ambulance I saw that the swampy end of the lake by which I was passing had a very strange appearance, so I pulled up, exclaiming to my men in astonishment, " Why, what are those ? " A steady look, and we each exclaimed, " No end of fowl ! " On this I gave the word to " hold on," or halt, and the next moment issued orders to Mr Canterall to pitch the camp.

This not being at all to the liking of that respectable individual, who was in terror lest delay should reveal to me his peculations, he replied " it was impossible to camp under the trees, as there was not a morsel of grass for the cattle." " Then go and buy corn," I rejoined, " which is better than any grass, and let me see the camp pitched at once, and no two ways about it." This command having been given in a tone which the fellow saw admitted of no dispute, taking my John Manton loaded with cartridge, I walked to within shot of some straggling wild ducks, and killed a very fine mallard, which Brutus brought me, and then what a cloud of whistling wings arose to fly further away to the broader portion of the lake ! Wild geese, ducks, and teal of every kind, all mingled in one confused flight, cackled, whistled, and winnowed through the air, and I retreated to my ambulance to get Pape's shot-gun, as well as my Manton, and some ammunition.

Having provided myself with these, and taken my faithful retriever Brutus with me, I started to go up one side of this extensive piece of water, but ere I could reach the woods on its forest bank I came to a corn-field wherein the settler was at work. Permission from him was readily obtained for me to go through his corn, when, having reached the woods beyond, I secreted myself in the limbs of a fallen tree, some of the bare points of which were submerged in the muddy water. By the time I had reached this my first ambush for fowl, the ducks and teal were still unsettled from the effects of my shot, and they kept on wheeling round either in large or small flocks or in pairs. A long but ineffectual shot at some wild geese alarmed them again, and then I had five long single shots at common duck, blue teal, and common teal, and Brutus brought them all to bag. A whistle then brought George to me, when sending him back with the fowl, I told him to order Tom to take his gun and go up the contrary side of the lake, to disturb the fowl from that side, to kill what he could, and to keep at it all day. Having given this order, George was, towards the afternoon, to follow me up the lake, and, keeping out of sight, by his ear to maintain himself within reach of a signal-whistle, in order to come up and carry home the spoil.

All this being arranged, I proceeded on through the wild forest very far up the lake, the fowl by this time having congregated in immense numbers. What a lovely, still, and sunny day it was for such a splendid prospect of sport, though the weather itself for that peculiar pursuit was unfavourable. The woods were so lonely, so ruinously grand, and so wildly beautiful in their picturesque decay, so hushed and so remote from man! It took me not long to discover that the lake was infinitely too extensive for me to do much with the fowl, for it could not have been less than two miles in length, and in some parts more than half a mile in breadth, the trees which fringed it having in many places fallen into its waters. Far from as well as near to the margin of its strand there looked up from beneath the surface the arms and points of old snags that had been submerged for years, now bleaching once more in the sun from the shrinking of the fluid in the summer heat. At first an appearance on some of these whitened limbs of trees puzzled me, for there were black-looking knobs on them, so they at first appeared to me, about the size in circumference of a dinner-plate, to which was attached something that resembled a long neck and head. These knobs exactly resembled turtle, and turtle they proved to be; they raised their heads in alarm at my approach, glided off the snags into the water, and I saw them no more. Innumerable frogs, who had been sunning themselves on the strand, took from beneath my feet the highest and largest and best " headders" into the water for frogs I ever saw, and showed me that the illustration of Dickens' Martin Chuzzlewit's seat in despair on the log in front of his doorless cabin in Eden had not been in any way overdone in regard to his reptile company.

Thus walking on the margin of the water, on reaching a little promontory formed by some fallen trees, I saw a few ducks sitting near enough to the bank for me to get a shot, could I but attain to the spot under the trees exactly opposite them. Having crawled on my hands and knees for some distance, to my great delight I saw a pig routing and knocking about the ground and bushes at the very place I desired to reach ; and this pig, having been seen and heard by the ducks, would lull their suspicion of other danger. I crept on, and induced the pig to see me when at some distance, in order not to occasion him any sort of surprise. At last the pig quietly walked out of my way, when, taking his place, I got a good shot at the ducks ; and Brutus brought me a couple of that lovely-plumaged bird the wood-drake, as plump and fine as it was possible to be. The noise of my gun having again scared the flocks of fowl, I crept on till I came to another hiding-place commanding a little bay. About the centre of the side of the lake, then, I sat down, close to the water, well screened by the huge limbs of a fallen tree, overgrown with creepers, and by my watch it was past noon. A more still and sunny prospect—the glittering though stagnant water fringed on both sides of its entire length, as previously described, by the autumn-tinted woods, and reflecting their varied hues—could never be viewed; while the music in the air made by the flocks of fowl calling to each other, or inviting their fellows to a descent to the bosom of the lake, was beautiful to the sportsman's ear.

Above me, at the opposite end of the lake to the one whence I came, the water seemed to be spanned by an embankment, which I afterwards heard was for a line of rail, and beyond this a fusilade of guns, as of a lot of skirmishers on a field of battle, had now become continuous; "hunters," as they call themselves, no doubt, blazing away at the fowl I had driven in that direction. Tom, too, by the report of his gun, I knew to be obeying my orders on the opposite shore, so I felt sure that between the one disturbance and the other if I kept concealed I must obtain some passing shots, so with one gun in my hand with loose shot, and the other lying by my side with the blue cartridge for long distances on the water (I would never use the green cartridge for anything but wolf or deer), and Brutus full of vigilance, but couched behind me, I awaited such shots as might come within the range of either charge. Very prettily, both on the water and in the air, was the sport accorded, the game varied by common duck, wood-duck, and the blue and common teal—the wild geese had taken their departure on the first alarm. While waiting thus, in the afternoon, three mallards and a duck of the common sort pitched into the lake just opposite me, but at so long a distance that I doubted the possibility, even with the cartridge, of a successful shot, so I forbore to fire, in the hope that they would come nearer. They were inclined to do so, but I think that after a time they detected some motion in Brutus's ears (it could have been no other stir of any sort), for they suddenly stretched their necks suspiciously, and remained quite still, their breasts almost touching each other. On this I resolved to risk the shot. Could I have measured it I am convinced it could not have been much under ninety yards, and on firing I killed the three mallards dead. In dashed Brutus, and when he had swum out to them he tried to get two of the birds into his mouth at the same time. Not being able to effect this, he brought first one and then the other, well knowing, as he brought me the first, that he left the other behind. As to the third bird, then, there arose some difficulty, for he had drifted away with a light wind that had arisen, and Brutus, not having taken any notice of him in retrieving the two others, came to the natural conclusion that this was all he was called on to do.

As fowl in myriads were still flying wildly about, I abstained from sending Brutus in to swim an immense distance he knew not whither or for what, and took up my position, again obtaining a few more shots, killing and losing some fowl from their being only winged at long shots, and too well able to dive. Tom, then, having rounded the lake by the railway embankment, came to me; he had bagged some quail by the side of the water, but his gun was non-effective at the fowl, though, to disturb them, he had tried long shots at flocks as he came

WILD FOWL SHOOTING.

along. He told me, in considerable mirth, when he saw what was lying by me, that there were two or three Americans firing away where I had heard them, but that they had failed to bring a single bird to bag. On this I bade him go back the same way he came, and tell them at camp to get ready my dinner, and that I should soon follow him. Having killed or knocked down another fowl or two, I then gave a whistle for George, who came up to carry home the game. We then consulted whether it would be possible to induce Brutus to comprehend that there yet was a dead mallard of the three killed at one shot far out upon the lake, and I resolved to try. Having spoken to the dog much as I would speak to a man, that " *Yes*, there was another bird *to fetch*," I signaled with my hand the direction in which I wished him to go, and the retriever plunged in, and after swimming a long way, looked back at me. I gave by my hand the signal of " on "—" away," and Brutus obeyed, till he caught a glimpse of the bird, and from an immense distance brought him proudly back to me, testifying his knowledge that he had reason to be proud, by walking around me several times before he would give the bird up. In returning to camp I met the settler who had given me leave to go through his corn, and presented him with a couple of ducks, when he very kindly put an empty cabin at my service if I chose to prolong my stay. Arrived at camp, I asked Mr Canterall if, with the day before him, he had got any milk, eggs, and butter, and received as usual the mendacious assurance that " he had tried in vain."

While I was at dinner, with my fifteen head of beautiful and varied fowl, as well as quail, slung on a cord from

22

tree to tree close to my ambulance, and three squirrels
which Martin had hit through the middle as usual with
his rifle, a nice, respectable, and very old man came
from a little cabin in sight, and, sitting down, entered
into conversation with me, and the more he talked the
better I liked his method and his manner. When he
arose to go, I asked him if he had any milk, as I saw a
cow or two about his door, and he replied that he had,
and two or three new-laid eggs, which he would gladly
offer to my gratuitous acceptance. Turning round to
Mr Canterall, I remarked " that he had taken much
trouble as usual to obey my orders," and then called to
one of the men to attend the settler to his cabin, and bring
back the milk and eggs; at the same time to carry with
him in return for the present a couple of my newly-killed
wild-fowl.

Having hastily eaten my dinner, I took Wallace with
me to the banks of the lake, the part nearest to my
camp that had been used by the wild geese, and, going
out as far as we could on the mud, I made Wallace bring
some boughs of trees to conceal me. We then thatched
ourselves and Brutus well over, and awaited the event. As
soon as it was dark, " Cackle, cackle! caw, caw!" came
the distant sounds of geese returning from the direction
in which I had seen them go—nearer and nearer they
came, but, alas! inclining to the contrary side of the lake.
I let them wheel round several times to ascertain their intention, and then becoming aware that they would approach no nearer, when they again came round I took a
long shot, but all in vain. After this it became pitch dark,
so, sending Wallace to walk up the lake for the chance of
his driving anything to me, Brutus and myself continued
our vigil. Nothing then could be more still than the night;

the stars were out, so that I could have seen geese had they come over my head, but not a sound of any sort could be heard, till, all at once, behind me in the forest, I heard the most melancholy noises, which I was sure were made by a large owl. While listening to them, again the beautiful wild cries of a considerable flock of geese came on from the distance, and they pitched in the lake out of shot, but just opposite me—so near, however, that in the stillness of the air I could hear them softly cackling to each other, and dipping and washing in the water. Wallace then returned, and got back to my ambush without frightening the geese, a proof of the darkness on the land in spite of the stars above, so I sent him again to get below the geese, and to clap his hands to scare them, for the chance of their wheeling over me. He did this, but failed in any effect, the geese quietly swimming away, and taking no further notice of the noise. We then returned to camp, and I " turned in " for the night.

The next morning, just at day-break, I went to the lake, for I could not devote another day to the fowl, and fired both barrels of my Manton with cartridges into an enormous flock of common teal who rose more than a hundred yards off. They were in such numbers, and so close together, that it was impossible not to hit some of them, and I knocked down five. Four of these escaped by diving out into the lake, but the fifth being killed dead, Brutus brought it to bag. We then commenced the march, my resolution being to trot on to St Joseph, as soon as I was sure of my route, to prepare things there for the reception of my establishment. Our way being through the woods and along the river all day, the road simply a track marked by a few waggon-wheels and the stumps of large trees cut just low enough to pass under the waggons, the

mules left to "straddle" the stumps or pass on either side.
I saw some partridges (which I did not stop to shoot),
but no other game, and at last got out of the woods into
a more beaten road.   We then came to where the road ran
in two directions, so I asked Mr Canterall which I was
to take ; he indicated a road with his usual effrontery, but
some time afterwards came up to say he had been wrong,
and that I must turn back.   At last I came to a very fair
road, and giving the rein to my mules I trotted on, and
not long after came in sight of St Joseph, on the Missouri
river, a straggling, but rising, and even then considerable
town, fast growing into importance, and on meeting a
citizen requested to be informed as to which was the best
hotel.

"The Planter's House, I reckon, yas, sir, 's as good as
any ; been on a hunt, I 'spose ?" eycing the fowl and
bison tails hanging around the front of my ambulance.
To this I replied that I had ; and very dusty, very hungry,
and very tired, I entered St Joseph and drove to the
Planter's Hotel, a building about the size of an English
wayside public-house.   Pulling up at the door, I went in
and found a free-and-easy young fellow lounging about
what in England would be called a bar; the floors of
the room, of course, dabbled all over with tobacco juice.
"Can you recommend me," I said, "to a stable-keeper
who can put up seven mules and four horses, my waggons,
&c., and can I have some dinner, a bed-room and private
sitting-room here for perhaps a week ?   I did not tell him
who I was, as I had found a man's assertion of his own identity was never believed in the United States, when I was
greeted word for word with the following reply.   "Guess
you *aire* too late for dinner (it was a little after four o'clock,
if I remember rightly), but you may have *a* room, *not*

two; what d'ye want with two, guess one's enough, ain't it?" "No," I said, rather riled, "it's my wish to have two, and I suppose in your free country I may have my pleasure in that way, and do as I like if I pay for it, can't I, and also have something to eat?" "Wall," replied this excellent limb of democracy or tyranny, not in purple but in rags, or of a man so mentally drunk with upside-down notions of his own freedom as to desire to make every other soul a slave—"Wall, guess you can have *something* to eat, and *a* room, and we'll see about the mules." "All right," I rejoined, on the eve of stepping nimbly out; "then just roast me one of my wild ducks for dinner." "Roast a duck!" said this upstart; "*no*, can't do that, its jest past our cooking hour, can't have anything hot." "Oh, I can't, can't I?" I replied, out of all patience, "then I will just exercise my freedom, and have nothing more to do with the Planter's House." The fellow stared at me for this unusual assertion of liberty, as if he could not understand my audacity, and I strode to my ambulance, and drove away, still a traveller in search of refreshment. In proceeding up a street I saw written on an hotel "Blackmore House." It was a dilapidated place, but still as good as any I had yet seen, so I pulled up, and having found the host himself, whose name heralded his hotel, I asked him if he could put me up and give me some dinner and *a* room. Mr Blackmore was all civility, and at once proffered me his best services and all the accommodation his house afforded, while at the same time he recommended me to an adjacent livery stable-keeper for all attention to my mules and horses.

"Lend a hand, then," I said to him and to his excellent black slave, "and please look sharp to my things." As I unloaded them there was a crowd of idlers soon assem-

bled, and while the black carried in my luggage I placed
the wild-fowl upon the raised steps of the threshold.   I
never left the place, and only took my eyes off the fowl
as I turned to my ambulance to unpack my things, but on
casting my eyes again over my birds I missed the only
couple of blue teal in my possession.   It was unlucky for
the thief that his taste for these delicious birds made him
select the only two of one sort that I had, for, of course,
without enumeration, I missed their blue wings, and at
once exclaimed to my landlord that some one had taken
two of my birds.   He said at first "They were all there
that I had put down;" but on my replying with considerable confidence "that I *must* know better, for that the
only two blue teal I had were gone," I saw him cast a
very brief look out of the corner of his eye at a young
man I subsequently knew as his son, and that youth instantly disappeared.   On this I dived into my ambulance
again as if going to the further end of it, but in reality,
when out of sight, I only faced about, when I beheld,
coming out of a side door of the building, a little way off,
the son of mine host carrying my blue teal behind him,
so I suddenly looked round the corner of the tilt of my
waggon, and said, "Oh! all right, you've got my birds;
I thank you;" when the young fellow, with a shabby
leer, like a man who had just picked his own pocket, deposited my birds along with the others, and said nothing.
"Liberty Hall here," I murmured to myself: "freedom
so much in the ascendant, I see, that each man makes free
with what in no way belongs to him.   All right!   I must
take it now that my things are unpacked, and, at all events,
I have one unusual comfort in America, I am about to be
robbed with civility.   Hooray!   I might meet with a
ruder thief if I went elsewhere."

In a short time I found myself ensconced in a dilapidated bed-room, and an object of considerable curiosity
as I ascended or descended the stairs to various misses,
young ladies who had come all without *chaperons* to
attend a ball to be held in the house that evening.
After dinner, at the suggestion of mine host, and a
gentleman or two that called on me, in my shooting
dress, for I had little else with me, I looked in at the
ball, and was introduced, as far as the very obscure
twilight of the apartment would permit, to two or three
people, but it was so dark I could scarcely recognise
their features. There were some fine girls there, as
far as outline went — the light admitted of no other
perception — and the dances indulged in seemed to me
to be made up of " quadrilles," " Pop goes the weasel,"
common English country dances, and " the Lancers,"
all pounded together, during the execution of which
complicated mysteries there was a good deal of loud
verbal direction as to the figure from the leader of the
band.   The friends with whom I had shaken hands
often suggested that I should join the dance, but as I
felt that that was utterly beyond the possibility of the
uninitiated successfully to do, I respectfully declined.
One fine girl having dropped a bow—I do not mean
that she knocked her partner down, but she lost a knot
or bow from her dress—I stepped forward, as all the
Americans seemed not to care about it, and rescuing
it from their feet, presented it to her.   She thanked
me, and the dance proceeded ; and some gentleman
standing by me said, " I had set his countrymen an
example in politeness."

CHAPTER XXI.

At St Joseph, as well as throughout society of whatever
rank in the United States, I invariably had found the
utmost civility and good fellowship, but in many of the
American papers that reached me, I suppose because the
" General Chokes " and " Jefferson Bricks " who edited
or owned them were personally opposed to peace—" war
pays best "—there was one continuous snarl at what
" the English Lord thought that he could do with our
biggest beasts." I had laughed, I never was riled, at
the trade or freedom in venom; but now and at this

moment I laugh still more when I read what the editor of the *Knoxville Whig* says of Attorney-General Mack, as quoted in illustration of the license of the American press by the *New York Tribune*.

The *Whig* says :—" We took a look at him (the Attorney-General), and we don't hesitate to say, that in his countenance we could see mingled the virtues of the Wandering Jew and the impenitent thief upon the cross! And if Attorney-General Mack is not a villain the Almighty does not write a legible hand, and this we are unwilling to concede." If neither the law in the United States, when guided by an Attorney-General, nor a revolver in a hand on the same side the Atlantic as the ruffian who can thus write, are insufficient to restrain this horrible licentiousness of the press, the offspring of an irresponsible freedom, why surely I can afford to laugh at less vile attacks, and to advise the nation to leave off the destruction of rattlesnakes, and raise a hunt to crush the adders of the Press that defile and render ridiculous the bosom in which they live.

Tired as I was, I did not stay long at the ball, but when I went to bed there was such a continuous " getting up-stairs," as it appeared to me, from the rooms below to those above, that my slumbers were somewhat broken. On the following morning I arose somewhat refreshed, and set about making arrangements for the disposal of my camping effects, my mules and horses. To this end I paid a visit to the livery-stables about the middle of the day, wherein also were deposited, with a wretched little room to themselves, George and my dogs. There was not a vestige of corn or hay in any of the mangers; the head of every animal I had was turned round at my approaching step, and on my going up to

Taymouth, Sylph, and my pony Charley, all of whom knew me so well, the rubbing of their heads against my arm, and their little suppressed neighings, informed me that they were very hungry. "When were my horses fed?" I asked one of the ostler Boh-boys. "Guess just now," was the reply. "That is odd," I rejoined, "for there is not a grain of corn in any of their mangers. What's the reason you do not let them have some hay to pick at?" "Reckon they won't eat it; 'tarnt used to it coming so slick off the plains." "I don't believe you have tried them," I replied, striding up to a quantity of dried switches called hay, and carrying a lot of it to Taymouth, who commenced eating it ravenously. "Now, then, young fellows, look alive, and just take the trouble to give my mules and horses the *chance of getting used to your fodder*, for it is my belief you have not done so yet; and if you don't pay more attention, I will report you all to your master, and if he don't make you attend to your duties, I will just make a clearance, and that in ten minutes, off your premises." This had the effect of filling all the racks with hay. The day for paying off my men having arrived, I soon got through this duty. The tongues of the fellows now becoming loosened, I quickly discovered what a thief Canterall had been; a dishonest as well as useless servant.

One morning before I had arisen from my bed I heard a gentle knock at my room-door, when on saying "Come in," my immense satisfaction and delight may be more easily imagined than narrated when the unexpected but much-wished-for apparition of my friend Mr R. Campbell, of St Louis, appeared, and acquainted me that he should stay in St Joseph for a day or two, on his way to Kansas city. A new face was soon put on my affairs, by his introducing me to a gentleman who would take charge of my camping effects and attend to their sale; to this gentleman I had already had a letter of introduction, kindly sent me by Mr H. Campbell, but which at the time of my arrival I could not find. From him, Mr Carbry, whose family I believe was of Irish descent, I received the most kind attention, as well as rather a superfluous caution, considering the experience I had had, not to place much faith in anybody. It gave me great pleasure then to lend my mare Sylph to Mr Campbell, who rode and was delighted with her, and being left more at liberty, my effects all on sale, I once more turned my attention to the adjacent wilds.

While at my window, or at the door in the street at night, I had more than once heard large flocks of wild geese in their flight along the adjacent river, and as at this time several sharp frosts occurred, though the sun at noonday was still intensely hot, I began to make inquiries as to the haunts of the various sorts of fowl. Having been called on by the mayor of St Joseph, who was to me a most kind, attentive, and hearty friend, Mr Jefferson Thompson, that gentleman invited me to a drive in his carriage, behind a very nice pair of chestnuts, to witness what in those regions is called a " barbicue." The occasion of this barbicue, or roasting of an ox and some sheep in the open air whole and for all comers, was to collect the people of the division of the State to vote a rate, as far as I understood it, to meet the expenses of a railway (the Flat river Railway, I believe) then in course of construction. Of course there were parties for and against the costly speculation, my friend, Mr Jeff. Thompson, being a leader in its favour.

After a considerable and hilly drive on a very cold day, we arrived at the straggling village where the ceremony was to be held, and where the polling-booth and speech-house were placed, and put the carriage into the hands of a black slave in whom my friend, the mayor, said he had perfect reliance; we then walked about, and I had full opportunity for observing that every decent person drove with bottles of brandy, cocktail, and other warm compounds in his pockets, or in the pockets of his carriage, and that it was the fashion of the times to exchange mouthfuls of strong liquors all day long. I declined compliance with this custom, founding refusal on national prejudice, and really my companion, in his good-natured way, taking my share perhaps as well as his own out of the best of motives, made me fear that I should have to drive him home instead of his doing so by me.

He more than once hinted that he wished me to see the rough and tumble onslaughts that would probably suggest themselves to these sons of freedom later in the day, but, as I assured him I cared for nothing but a fair upstanding fight, and that I would prefer going home in peace to the chance of having to lick somebody or being roughed and tumbled myself, he stood manfully against the many persuasions that were made to him not to go home before night, and having seen all that was worth seeing early in the day, we prepared to make a start on our return to St Joseph. While at the barbicue I went into the speech-house, to hear the mayor discourse to the people, and therein saw a novel and effective appendage to the public rostrum, and one which I would suggest for the Speaker's adoption in place of the table at which the clerks sit in the English House of Commons. Immediately beneath the

platform on which a speaker stood to address the enlightened men of freedom in the body of the hall, at a table to
hold their music-books sat a numerically-sufficient band,
of course accompanied by that bully of all rurally-played
instruments, the big drum. This band was *piano* even
to entire silence, or *forte* to the effect of drowning any
other noise, precisely as the majority of the mob in
the body of the building assembled called out to " Hear
him," or "Music." I can assure my readers, that this
excessively effective and *recherché* method cut short every
"sad tale, saddening because 'twas doubly long," and
rendered useless the adoption by our, in this instance,
more refined transatlantic brethren of the horribly vulgar
noises, caterwauling, cockcrowing, and "oh! ohs!" with
which the members of the English House of Commons
seek to put down tedious speakers. I had not an opportunity of addressing the Americans on this platform, or I
should certainly have complimented them on having
whipped the mother country by chalks in this harmonious
method of obtaining silence.

In our way back to St Joseph, in passing a bevy of
coloured men (slaves), the mayor said, " Now you shall
see the men who really enjoy themselves; they are on
holiday to-day." Thus saying, he pulled up, and, addressing the blacks, asked them who could " pat " I think
he called it, as an accompaniment to a dance and song.
For some time these men of colour were bashful, but, at
last, a woolly-headed really free man—if freedom from all
care of self-maintenance constitutes it—commenced patting his thighs, singing, and dancing; when this being too
much for his brother blacks, in a short time they were
worked up to the blithest pitch of merriment and activi-
ty; and when we had seen enough of it, the mayor tossed

among them several of the smaller coins, for which there was a general scramble, and we drove away.

We had not proceeded far when we overtook a small boy with a long rifle, at the sight of whom the mayor, with good-humoured exultation, exclaimed, "*Now* I'll show you a specimen of how we shoot in this country. That little fellow, trained to the rifle from his cradle, would hit the pip in the ace of hearts, as far almost as he could see it." "Come here, boy," he cried, jumping out of the carriage, and producing from his pocket a penknife and visiting card, sticking the latter, by the aid of the former, on the bark of a large tree by the road side. "Now, boy," taking him back the very short distance of twenty paces, "show this gentleman how you can shoot. If you hit the card I will give you 'a bit,' but if you hit the penknife you shall have a dollar." On this the boy grinned with delight; but as my friend the mayor had stuck the card on a tree exactly in a line with, but the road wide, from me, I begged the shooter to pause till I attained a safer place, in case the ball should glance, the mayor ridiculing my apprehension. Having allowed a more considerable margin for accident or juvenile misdirection, I watched the shot, and to my intense amusement this young example rifleman, after some moments' aim, not only did not hit the knife or card, but he missed even the tree! The mayor himself, though considerably disappointed, laughed at the failure, resumed his seat in the carriage, and we had a very jolly drive the rest of the way home, meeting a gig or two, containing some friends of the mayor, who, like highwaymen of old, stopped us and produced their pocket pistols at our heads, demanding, not our money, but our brains, the power over which, and in regard to myself, I resolutely refused to resign.

While I remained at St Joseph, I was introduced to
some very nice young men, high-spirited, and gentleman-
like withal, as well as to an old frontier huntsman, Mr
Davis, who was supposed to know the haunts of every wild
thing in that portion of the United States. Mr Davis,
according to the information I received, was one of those
many instances of being as it were overtaken by good
fortune, even after the recipient had made enough to supply every moderate want, and had sat himself down,
milestone fashion, by the side of the highway of the
world, a thing which so frequently happens in the sudden
growth of that gigantic country.

When " St Joe," as the town is familiarly called, was
in its infancy, Mr Davis had purchased some land, on
which he resided, at a little distance from it, taking occasional trips into the desert, when one, I conclude, very
fine morning, a customer on a large scale walked in,
foreseeing that the town was going out of town, and bid
for and bought the land at, so to speak, an immense value.
Mr Davis consequently again purchased land a little further out, and with his unexpected wealth, not in the least
wanting such a residence, built himself a good house, and
again sat down, according to my belief from what I saw
of the prosperity of St Joseph, if not in his life, still in
that of his immediate successor, to be once more sought
by gold.

On the first Saturday after my arrival at St Joe, Mr
Davis having been appealed to as an oracle, Mr Shields
(one of those agreeable acquaintances it was my good fortune to meet with) and myself agreed to accompany the
old frontier man on a sporting excursion. The spot selected for " the hunt " was about eight miles' distance
from the town, and thither we resolved to proceed in my

ambulance, behind my favourite mules, accompanied by
my retriever Brutus, as Mr Davis assured me we might
shoot our guns red-hot at geese, ducks, and every sort of
fowl. Alas! I had then become too much used to the inflamed minds of Americans to put implicit faith in any
story or asserted oracle in the shape of man, and for that
reason I encumbered myself not with an overweight of
ammunition. However, in case there should have been
a demand for it, in my ambulance I put some spare cartridges for fowl; but they were in no way needed. On
meeting in the morning for our start, in the hands of Mr
Shields was a handy double-shot gun, rather small and
not in good order, but on the shoulder of Mr Davis there
indeed was a powerful heavy double gun, beautifully kept,
and of a size which in England would have been exclusively set aside for geese and ducks, and which, so its
owner told me, was of English make.

We had a rough drive, but at last came to a wide
stretch of grassy, and perhaps in winter of marshy land,
intersected with woods, and in the vicinity of the Flat
river. On this plain were a good many small ponds of
water full of reeds and rushes, and in the woods a cabin, at
which we halted the ambulance and fed the mules. For
a huntsman of the frontiers, famed in those parts for his
knowledge of woodcraft and water-fowl, I could not help
being surprised at Mr Davis's dress. He, like the fellow
called my guide, was perpetually cumbered with a bright
sky-blue great-coat, so that when miles off on the plain
he was for ever visible, resembling the moving egg of an
English hedge-sparrow; and, more, three other sportsmen whom I met, all riding about the plains in quest of
something, but without dogs, were every one of them attired after the same fashion. As I understood that these

were all military great-coats, from the Government stores, some one, or the nation generally, asserting its freedom in this as in all other things, had, like Falstaff, "misused the king's press damnably," for the blue coats were in nowise patched nor worn out, but in a state of very serviceable perfection.

On starting for our hunt, the old frontier oracle always strayed off alone, leaving Mr Shields and myself together, and I suspect for the fair purpose of contrasting the bag he expected to make against mine, though I stood to him, by the presence of Mr Shields, in the light of two to one. As I anticipated, there was not such a thing as a wild goose to be seen, and scarcely a duck. On coming together early in the day, brought so by the report of Davis's gun, Brutus had to give that old hunter a specimen of the craft of an English retriever, for Davis had killed a wood duck, and it had dropped in the water in the midst of the arms of a large and fallen tree. The bird was brought to bag. After this little episode we did not again meet till the day was done. While beating for game, Mr Shields and myself came upon a large pond in the woods, completely filled with rushes; when encouraging Brutus to go in and hunt it, a small flock of English wild duck arose, and we fired three or four barrels into them. Two ducks appeared to fall dead, and a couple more fell at some little distance. Those that seemed to fall dead, however, were only winged, when, from the extent and depth of the water, and thickness of the rushes that grew in it, though Brutus worked for an hour or more, we failed to get them. The two that fell some distance away were quite dead, and these we obtained.

23

While endeavouring to recover these winged ducks,
three blacks (slaves), who were cutting wood, came to
stare at us, so I offered them half a dollar each if they
would go into the water, and stand in it at given points
to help the dog; but they shook their woolly heads in
horror, and said they would not go in there for any
amount of dollars.   I was in the water up to my knees,
or as far as my boots would let me go, and I am sure no
part of the pond would have reached over the waist of a
short man.   On receiving the refusal to go into the
water from the spokesman of the slaves, with the addition that he would not go in for ten dollars, I replied,
"Hoorah for the slave, then; he's not in want of
money, however he may be of soap to wash him
white!"—this sally of mine was received by a shout of
laughter from the whites at the wood cart who were
looking on.

Having toiled in vain after these ducks, I sat down on
the bank of the pond to refresh myself with an excellent
apple given me by my friend Davis, and while thus
employed asked Mr Shields what those things, grass-
cocks or little mounds of grass and rushes, were at intervals all over the pond.   After some hesitation he replied
"that they were made by large ants."   "That can't
be," I rejoined; "for I know of no water ant, nor any
insect that would make such places in the middle of
ponds.   They are made by some animal, and I suspect
the musk-rat."   As to this my friend, with that little
notice of the natural history of their own country taken
by Americans, could give me no certain reply; but on
joining Mr Davis, he confirmed my suspicions, and told
me that the Indians would often, out of one of these

mounds of grass, spear several musk-rats, which served them as an occasional article of food.

Late in the day Mr Shields and myself fell in with two or three coveys of quails, but, as usual after the first rise, they took to very severe cover. When they were scattered, he imitated the call of the old bird to perfection, and invariably obtained an answer, which enabled us to have some sport. On meeting at my ambulance when the day was done, the entire bag consisted of nine quails or partridges, a couple of common wild ducks, a wood duck, a couple of the English whole snipe, and a squirrel. Of these the old frontier man to his share obtained but the wood-duck, the squirrel, and, I believe, one quail. Mr Shields, in addition to his being a most agreeable companion, was a very neat and quick shot at quail, but the hammer of one of his barrels got wrong, which, of course, baffled him occasionally. Davis never let me see him shoot, so of his proficiency I am unable to speak; but in kind attention and good humour in those particulars, he vied with all my other friends in St Joseph. My companions pretty well tired, we drove home by a rougher road still. My mules jumped some bad rills of water, and dragged the front wheels into the mud and against the opposing bank up to the axles with such a jerk as would have smashed any vehicle of English manufacture ten times over. During the day I had a view of a large bird of prey, but too far for a shot, which Mr Shields told me was the bald-headed eagle. From a dark night, and a darker road, we were at last delivered to the still more undefined streets of the town, intersecting the independently situated houses through which I worked my way with the reins, Mr

Davis the compass by which I steered; and we separated at mine inn, when, I much regret to say that, from wet feet and the cold drive home, Mr Shields was stricken with his periodical return of fever and ague. For myself, I ate a very hearty dinner and slept till daybreak the next morning.

In conversation with my friends at St Joseph, I found that the distemper in the United States, and particularly in their vicinity, was very fatal to dogs, and that insanity arising from the virulence of that disease frequently, as in England, was mistaken for that surely fatal malady, the hydrophobia. It gave me much pleasure to impart to my friends all the knowledge which my experience afforded as to these matters, and I hope that by so doing many dogs in America, for the future, will not only be more scientifically treated, but that many a human being will be saved the misery of apprehension.

It soon became manifest to me that all customers or purchasers for my camping effects and stock were holding back under the knowledge that I was in haste to be gone, and, therefore, that at the last moment I should probably sell everything for an old song. However, on receiving a bid of 115 dollars for my ambulance waggon, and a few camp-fixings, I took it, though that sum was not half its original cost. I knew too well the marketable value of second-hand carriages, and the rough usage the springs and wheels had received, to stand out for a higher price, and myself and my customer were mutually pleased. My young retriever Alice now had her puppies, but they died soon after birth, I think from not having a sufficiently soft and warm bed. These puppies were a loss to me, as I could have sold them with their mother at a high price, my mind having been made up to let her go, from a growing inclination which I had perceived in her to severely bite her game. She was not one of my best sort, so the parting would have affected me but little. My mules and horses, though standing at "six bits" a head, did not improve in their condition, and I strongly suspected a design to weary me into a hurried sale, a thing which I tacitly determined not to submit to, but, on the contrary, to give the dollar-seeking men by whom I was surrounded a considerable disappointment.

On Tuesday, the 31st of October, myself and two friends, Messrs Shields and Baxter, fixed to shoot on the ground immediately adjacent to St Joseph, but on the opposite side the river, and for this purpose we took with us Brutus and Chance. Two better or more agreeable companions than these gentlemen were could not be, when, as there were plenty of partridges or quails, I enjoyed the day very much, though Mr Shields still suffered from fever and ague. The covert (these birds when disturbed always fly to covert) was very severe, dry, and dusty, and the sun intensely hot, but between twelve and half-past four we bagged seventeen brace of partridges and three prairie grouse. We also killed, as we crossed the river at starting, a duck, which resembled the tufted duck in England, but was not quite the same bird. Both Chance and Brutus were the admiration of my friends. During the day's sport, though these gentlemen shot very well, they acknowledged the superior shooting of my favourite John Manton gun, and again I had to remark that the American sportsman never attempted a long shot, and since I have tried the powder manufactured in the United States I do not wonder at it, for it has not, taking the average of it, half the strength of the English powder.

On the following day, the 1st of November, with con-

siderable amusement I turned the tables on those expect-
ant dealers who were lying in wait for the hurried sale of
my effects, by starting horses, mules, and dogs off by a
steamer, the Carrier, under the charge of that excellent
gentleman, Capt. Bailey, direct for St Louis, insuring
the two former at 1400 dollars, and consigning them to
my friends, Messrs Campbell and Co. Here, again, I
found that dogs, however valuable to their owner they
might be, could not be insured; so, while we can neither
travel our dogs by rail in America, other than through
the permission of the baggage-master, and under the
miserable extortions that class of individuals chose to in-
flict, nor by steam on the dangerous rivers in insured
safety, nor under laws prohibiting theft, I advise my
brother sportsmen seeking the Far West to leave the dear
companions of their leisure hours behind them, and to be
contented in regard to all the smaller game with a much
less bag than the English dogs would have procured.

It was about this time that, while at St Joseph, I received the following letter:

"St Joseph, Missouri, U. S. A., Oct. 31, 1859.

"Honourable Sir,—Appreciating your high position
and ability to form a correct and intelligent opinion of the
people, manners, and things that pass under your observation, we would be pleased if you would favour us with
an opportunity of hearing the impressions which our country has made upon your mind. If, therefore, your time
and inclination will permit, we would be happy if you
would appoint an hour at which it will be convenient to
meet us, that our fellow-citizens may hear your remarks.

(Signed), M. Jeff. Thompson (Mayor), Jas. Craig,
    Nillard P. Hall, F. W. Smith, John Carbury,
    George Vanden, B. M. Hughes."

Having at once consented to give a lecture, on Wednesday, the 2nd of November, the large room taken for that purpose by the gentlemen of the St Joseph Institute was filled not only in its seats, but in every portion of its standing room; and on ascending to the table placed for me, I cast my eyes over as nice and pretty an audience as any public speaker need desire. There were in the front seats a very large preponderance of ladies—a fact affording to me very considerable satisfaction, as hitherto my converse had been entirely on the male side, and I was of course chivalrously desirous, after forming so good an opinion of their protectors, to see the beauty that must have had its customary effect in rendering the men so well worthy to be associates in any polished society. Nothing could have been more kind than the reception given me, nor better or more gratifying than the grace and heartiness of the applause. In my address I spared neither fault nor foible that had come across me in my travels, nor did I forego my praise and admiration when I deemed them due; and, among other things, the state of the laws, or rather lawless and murderous misrule which governed the American duel, fell under the lash of animadversion.

After the lecture I accompanied some gentlemen to their store, where I joined them in a glass of brandy-and-water, and then paid a visit to the library and reading-rooms of the St Joseph Institute. The members of the society received me there with the utmost good fellow-ship and cordiality, and I subscribed my name to their association, the first of my countrymen, though I trust not the last, who will set his hand to a list, in my opinion calculated to do honour to a rising, and, one day to be, a most important city. On the following morning a de-

putation from the young men of this society—young men
of whom I am sure that I can safely say that every soul
of them would in a moment have risked his life in the
cause of honour or duty—waited on me with a vote of
thanks for placing the true interests of an appeal to arms
in their proper light, and combining an assurance that for
the future, and where any of them were concerned, the
strict rules of honour, and all just avoidance of unnecessary
bloodshed in a duel, should be the chief object sought;
and that, as in all cases one must be wrong, the barbar-
isms and uselessly sanguinary idea of " the duel to the
death " should be discarded.   Would that every approved
English gentleman who may succeed me in a visit to the
plains would still impress this salutary and Christian advice upon the gallant men of America, for they need but a
resolute example from a known and approved man to lead
their high spirit in the true direction.

Mr Carbury having kindly taken charge of the rest of
my camping effects, I now took leave of my friends, and
proposed to take the rail at six o'clock in the morning—
thus, by way of Hanibal, to reach St Louis before my
mules and horses.   The night before I left, some of my
friends came to see me, and I am sure we enjoyed the
whole time we were together.   As a faithful historian, I
cannot but here relate things of some of my countrymen
whose birth or station in life ought to have taught them
more caution; their conduct was thus narrated to me.   In
the Far West the English gentleman of course is little
known, and therefore those who go there ought to be
doubly sedulous to create a good impression.   Some of
my friends in St Joseph then told me that two of my
countrymen who had been there received the visit of my
informants to drink some wine with them, seated on the

floor of the room, without their coats, all chairs and tables removed, and the bottles and glasses simply at hand, and in that position and after that fashion they attempted to make their visitors intoxicated. " What was the event ? " I asked, pretty well guessing that men who could drink drams and gin-and-brandy cocktail in sips all day would be difficult customers to give a wine quietus to in a short night. " The event ? " replied my friends ; " guess we left your countrymen on their backs, and walked home as steady as a statue in a square at New York." " Were you not surprised," I continued, "at the method of your reception, or did you think that that was the way an English gentleman in his own country would behave ? " " Guess we pretty well knew it was poking fun at us, and thought they, your countrymen, were greater fools than we were to suppose that we did not know your customs better." " Quite right," I replied ; " and now let me advise you, my friends, for the future to deem the following a test of the real English gentleman :—If any Englishman, Irishman, or Scotchman comes here, and calls himself a lord or a gentleman (the terms ought to be synonymous), and puts on any sort of affectation, be quite sure that he is an impostor, and not a leader in the best society of the Old World. Affectation or any pretence at eccentricity is the height of vulgarity, and neither in a foreign land, nor anywhere else, will a true gentleman forget himself in such miserable assumptions."

On the morning of quitting St Joseph I was called at half-past four, and at 6 a.m., having had no breakfast, the rail bore me off in the direction of Hanibal, distant 206 miles, and alas! afforded no time for eating till twelve at noon, when we stopped at Brookfield and had time enough to obtain that which to me was a hasty breakfast, but which to others was a dinner. At Hanibal we took the steamboat for St Louis, which was not unpleasantly filled with passengers, and I had again to admire this method of transit as compared with the rail. On board the best boats of the Mississippi and Missouri rivers good order is maintained, the captains are intelligent and civil, and there a man need not sit by such filthy companions as are forced on him in the railway carriage.

CHAPTER XXII.

DURING the voyage from Hanibal to St Louis I had a good deal of conversation with two Americans who had been angling, and who had with them a very good basket of fish, consisting of the freshwater bass (much resembling our perch, but not of such fine hues), which they told me would often run as heavy as 8lb. They had also the crappè, so pronounced, which was very like our freshwater bream, only not so large, and a freshwater dog-fish, a nasty-looking thing, something like a *Salmo ferox* out of season—as coarse a commodity as ever was covered with scales. From all these I took exact sketches, which I have given to my friend Mr Francis Francis. There are two sorts of perch, the black and the white, the former of which, though not so good for the table, rises readily at a fly. For bass, perch, and crappè, the bait generally used is a small chub, spun as we spin the minnow for the same sort of fishing; for large chub, the grasshopper, or worm, is a very good bait. There is also another fish called the "yar," a thin-shouldered fish, with long mouth, and not very good eating. In the back streams and tributaries to the Mississippi, in the spring and fall of the year, excellent sport may be had with these fish; and I have no doubt but that good bright artificial minnows, or decoys of any kind, would be very killing; but then, from the innumerable "snags," so many sets of hooks would be lost that it would be an expensive proceeding.

On this voyage I again met a countryman of mine, Mr Shaw of Kentucky county, who seemed not to be able to show me enough attention. Among other kindnesses I received at his hands was a bottle of first-rate American whiskey, in the sale of which he was professionally interested. On board the steamer were some "Pikespeakers" returning from the gold-fields, and in their possession were the antlers of two wapeties, who they, with the customary misnomenclature of their countrymen in such matters, termed elks' horns. The finest of the two heads, with the aid of my friend Shaw, I purchased for fifteen dollars. Having made this purchase, I carried the horns into the saloon and put them with my luggage, and they became a matter of much interest to the cabin passengers. Here, again, American inquisitiveness, and, as regards an English gentleman, mistaken estimation of character, was most amusingly displayed. A fellow followed me up from the deck where I had made my purchase, and who had been present when I made it, and mingled himself with the crowd in the cabin, looking at the horns. Presently the sneak came forward, and, I suppose, judging me by himself, he, though he had seen me purchase the horns not a moment before, asked me " to give some account of the great sport I must have had in killing so fine an animal." Looking him full in the face, with all the contempt I could muster, I replied, " You know very well all about all the sport those horns gave me, for you were present when I bought them." The fellow slunk away disappointed.

At half-past nine a.m. we reached St Louis; and, not liking the attention I had received on my first visit at the Planter's House, I went to Barnum's Hotel, and on arriving there was charged by the flyman two dollars for simply conveying me from the quay hard by. I gave him one, and told him I was too wide awake to be imposed on. The captain on board the steamer had been, as usual, all I could wish; and, having secured a double berth exclusively to myself, I had been comfortably off. The black barber on board cut my hair, and officiated on my head with a beautifully cooling wash, and a particular way they have of gently kneading the skin, that was perfection. Douglas's Promethean Balm, from Old Bond-street, which I had with me, and which I infinitely preferred to any meretricious pomatum in the possession of the sable artificer, immensely charmed the knight of the comb and scissors, and he said, " Sar, yes sar, with this the baldest customer I have would soon get up a head as hairy as a buffalo!" " I fear not," I replied, " for I always use it, and my hair is getting thin." What I should have done without my tin bath I know not, for in this steamer there was not even a basin attached to a berth; but if anybody thought of washing their hands, they had to repair to two or three public basins set apart for the purpose. The anglers whom I had met on board the steamer kindly presented me with the largest bass or perch, and the crappè, and they were excellent at dinner. Oh! what luxury there was in sitting down to a well-dressed dinner at Barnum's Hotel, civilly cared for in every way, and everything looked to by a responsible steward. The waiting and the dinner were excellent, and I soon found that in this my second visit I had gone to the right hotel.

The next morning, on ringing my bed-room bell, it was answered by a very small boy, who entered the room just as if I had invited him to a game at marbles. "Get me some hot water," I said. "Can't," was the offended reply; "I'm not the hot-water boy." "What are you, then, young fellow?" "I'm bell-boy." "Well, of what earthly use are you?" "Do n't know." "Then return to whence you came;" and as he retired I again rang the bell. To this, my second summons, a larger boy appeared, who on being asked for hot water, asked in return, "Where's your jug; you're just from the plains, arn't you?" "What is that to you?" I replied; "and how should I know where the jug is? Go and get one, and be quick about it, or I will tell Mr Barnum." This threat had the desired effect, and presently after the same youthful limb of democracy re-entered, bearing in his hand a small pitcher of tepid water, which I threw into my bath, and peremptorily ordered him to fetch me some that was boiling hot, as I wanted it for the purposes of shaving. He was not long before he

came back again, when skipping nimbly across the room
he hastily set down a jug, and, blowing his scalded
fingers, exclaimed with much emphasis, "Guess I've
fixed you this time." I laughed, and so did he. On
looking at the trousers I had pulled off, I observed that
the buttons on the strap at the foot were gone, so again
I had recourse to the bell. A third youth, bigger than
the others, came, and having been made aware of my
desire to have buttons sewn on, he departed in possession
of the raiment, but very soon returned, and told me,
" Gentlemen and ladies down-stairs can't do it nohow;
but to oblige me he would take it out by-and-by, and
get it fixed elsewhere." "Upon my soul, you *are* oblig-
ingly handy chaps here; perhaps I had better do it myself," I cried; which seeming to interest him much, he
laughed, shook his head, and again vanished, materials
in hand. He had not been gone many minutes, when
he returned with a much more civil manner, begging
my pardon, and assuring me "that the housekeeper,
who had been out, was returned, and had been very
happy to sew on the buttons."

On the voyage from Hanibal to St Louis I had seen
those huge rafts of timber, piloted by from eleven to
fourteen men, with six or eight rudders to them fore and
aft on either side, and again had to remark on the unwilling mixture of the waters at the junction of the
Missouri and the Mississippi. For miles, though flowing
together, the waters are as different from each other as
those of a stagnant pond from the ocean waves. The
voyage during the night, beneath a bright moon, had
been very beautiful, and the contrast of the silvery and
placid light in the sky with the angry glow and unsteady flare of two large fires, one on the prairies, and

the other in the woods, was deeply interesting; as the
air from the fire in the woods reached my cheek, it came
with a warmth that told of the terrible heat over which it
had passed on its way to the water. On board the steamer,
as usual, there was plenty of ice and sherry cobbler.

On Saturday, the 5th of November, a boy came up
to my room with a card that had been left for me, with
a message on it obscurely written. When he entered I
was reading a newspaper which had arrived from
England, so, on obtaining the card, I put my paper
on to an adjoining arm-chair close at my side, and
began to try to make out the intelligence. This
being a difficult thing to do, I became deeply engaged
in the study, when, on hearing a noise in the adjoining
arm-chair, I looked round and perceived that the lad
had coolly thrown himself into it, and, cocking up his
legs over the side next me, had become busied in
the perusal of my paper! On this I snatched it out
of his hand, and bade him "stand up and prepare
his empty head for any message I might have to give."
He obeyed me, but it was with a look of such surprise
and burlesque of offended dignity, that I burst out
laughing; he never attempted, however, to reseat himself in my presence.

On Monday, the 7th of November, my baggage had
not yet arrived, and it was this day the committee of
managers of the public library and lecture-room called
on me to prefer a request that I would give a lecture
there, which I agreed to do on the following Wednesday
evening. On Wednesday, the 9th of November, on
repairing to the rooms, I had my four favourite dogs,
with their chains and collars, tied up on the four corners of the platform, and on either side the desk at

which I was to speak. While I was in an adjoining
room awaiting the hour of performance, I was surprised
by hearing a round of applause from the guests already
seated. This was occasioned by Druid, who rose from a
slumber in which he had previously been curled up, and,
with his long ears and sagacious face turned in surprise
to the body of the room, steadily regarded the ladies
and gentlemen. The gravity with which the old bloodhound did this, as well as his remarkable, and, in that
country, unusual appearance, I am told was irresistible,
and hence the burst of recognition.

When the time came for my appearance, nothing could
be more flattering than my reception from more than a
thousand people—I believe the room will hold seventeen
hundred; but though the night was terribly wet and cold,
there was no such thing as a moderately-charged-for cab or
fly to be had, those vehicles, as I have before stated, resembling small Lord Mayor's coaches, and only to be hired
for a price utterly beyond the reach of any but a rich man.
In consequence of this, many ladies were kept away; but,
as it was, the gentlemen assured me that it was the largest
attendance that had ever patronised a lecture. In my
address I again spared neither fault nor foible that had
come within my notice while in the United States, nor
did I abstain from the gratification it afforded me of
bestowing ample praise on all that demanded such
acknowledgment. The entire lecture, lasting a little over
an hour, was well received, and at its conclusion a friend
of mine told me that a gallant but retired officer of the
United States army assured him that he went to hear the
lecture in such low spirits, and under such a melancholy
feeling from adverse circumstances, that at the time he

thought he should never smile again.   However, the result
proved that his anticipations were wrong, for at the conclusion of the lecture he found that he had been laughing till
his eyes ran over with mirth for more than an hour, and
that, too, at the faults and foibles of his countrymen.   The
moment I concluded my address, in the midst of the kind
applause bestowed upon me, the men from the body of
the room rushed at the platform, took it by storm, and, to
my anxiety, for I expected to see Brutus, Druid, and Bar
tearing at their legs, vied with each other in shaking me
by the hand.   The utter astonishment of my faithful dogs
was beyond description : they were enraged but mute and
motionless, and lost, as it were, among the many legs, on
which limb to wreak their vengeance.

On Friday, the 11th, having made one of the most
agreeable acquaintances of my life—that of Dr Pope—I
accompanied that gentleman over his lecture-hall and
private anatomical museum,—as a private collection certainly second to none in the United States nor anywhere
else,—and received from him several curious presents in
natural history, as well as the most perfect little *bijou* of a
revolver I ever saw in my life.   The makers of this perfect
little weapon, which is highly finished and handsome
enough to be worn on a chain at a lady's waist, are Gt
Sharp and Co., Philadelphia, and for which in 1852 they
had a patent.   In size it is so small that I carried it in my
waistcoat-pocket, and in execution so effective that at eight
yards I could shoot as correctly, if not more so, than I could
with my favourite pair of John Manton duelling-pistols,
as they used to be called, and the little conical ball propelled by cartridge, at ten yards would go through a
half-inch deal board.   This little pistol is a four-barrel
revolver, loading at the breech, but revolving in the

hammer, each cock of the hammer bringing into its proper place the right nipple or peg for concussion on the respective cartridges.   I have been told (I never saw it so used) that this deadly little weapon is made to come into play in those brutal and bloody " difficulties," as they call them —they cannot, indeed, be called duels—which, I regret to say, so frequently disgrace society in the United States. Supposing one free citizen to deem himself free to take the life of another, and to enter at once, without counsel or friendly and cool advice, into a " shooting difficulty," he puts his hand, containing a six-barrel revolver, into his bosom, cocked and ready for use, and walks out for the purpose of meeting his victim.

He sees his unconscious victim approaching, and, suddenly confronting him, revolver, though concealed in his hand, ready for momentary action, in order to ease his shadow of conscience, he calls his victim a liar, and bids him draw and defend himself.   Supposing that the unprepared victim does not possess Sharp's little " pocket compeller," but has in a pocket a revolver, if, on the word to draw, his hand attempts to do the office, his bloody antagonist shoots him dead before he can touch his weapon. If, on the contrary, the supposed victim is duly prepared for any emergency, and happens to have in his pocket " Sharp's compeller," he temporises with his foe somewhat in this fashion : " Guess, neighbour, you 're mighty sharp on me, you air ; I do n't want none o' this.   Carn't we better fix it no how?   Jest you suppose now, if 't warn't be best for you and I to see if we can't call in a mediator ?"   All this time the intended but unwilling-to-be victim is apparently only fumbling nervously in the pocket of his coat, but really he is bringing the muzzle of his little compeller into the front corner of his pocket, and levelling that pocket corner, which of course his downcast eye can see, at the fourth or fifth button, as the case may be, of his adversary's waistcoat. Having perfected his aim while temporising in speech, he shoots his opponent through the heart, lungs, or body, no matter which, and effectually unsteadies his neighbour's aim. Society at large in the United States is much indebted to Sharp and Co.'s patent, for the civilised humanity of such a weapon is obvious, and needs no further praise from me.

My horses, mules, and dogs having safely arrived at St Louis, and received the kindest attention from the captain of the Carrier steam-packet, I resolved to leave all but my dogs in the care of my friend Mr Campbell, to be by him disposed of. George and my dogs I despatched by express, and in kennels or roomy boxes made for the latter, right through to New York, and from Adams's Express Company and their servants I obtained all the prompt notice and care I could desire; nor must I omit thus publicly to thank My Hoey, their gentleman-like agent, for his never-failing kindness and attention.

On Sunday, the 13th of November, I attended in the pew of my gallant friend Col. Sumner, at Christchurch Church, and therein heard, from the Rev. Mr Schayler, as gracefully preached and as effective a sermon as ever man uttered. The address which I then heard delivered before that attentive and respectable though scanty congregation thoroughly confirmed every idea I had previously formed of the state of religion generally throughout the United States. True, at New York and in other civilised cities fashion makes the church to be well attended, but the further you go from fashion and the best society, and the more you get among the middle and lower classes, the greater the neglect of all spiritual consideration. Step by step, from New York to the desert, I observed this, and it was that which made me previously say that among the last settlers on the frontiers tobacco and whiskey went ahead with man precisely as religion and roads were left behind. To such an extent does this oblivion of the Sabbath-day go, that from the want of one day of rest to distinguish from the other six days not one man in ten of the Far West settlers can tell you if you ask him the day of the week. All days are alike, and not one of them is set apart for rest and worship. At Christ-church it was very remarkable the total absence from divine worship of what would in England be the labour-ing classes, and hard indeed did Mr Schayler hit the irre-ligious feeling of the times, as well as that of the larger portion of the inhabitants of the city wherein he preach-ed, when he instanced the fact that "in periods of idolatry it was easier to find many gods than one honest man, but that now and in St Louis and among certain classes if hon-esty, however scantily, existed, the quotation, numerically speaking, was on the side of men, for it was difficult to find one god."

In St Louis there are an immense number of Germans, who drink, dance, and sing disreputable songs all day, and in this country, intoxicated as it is with a freedom, extending from the press to all the lower classes of society, and amounting to sheer licentiousness, there is not a governing power, through police or executive law, to check this foreign importation of sin from mingling and inflaming the vicious propensities of the people inherent in irresponsible numbers, and rife from an unbelief in heaven. The orchestral service was very well performed at Christchurch, but there was no one to officiate as clerk. During this my second visit to St Louis I made the

agreeable and instructive acquaintance also of the Roman
Catholic priest, the Rev. J. De Smet, who kindly presented
me to Mrs Tucker, the superioress of the Convent of the
Ladies of the Sacred Heart, the convent founded, if I recollect rightly, about the year 1830.   I visited in company
with my reverend friend the Roman Catholic college and
convent, and nothing could be better or more perfect than
the order in which they were kept. The dormitory in
which I met the young ladies assembled at the convent
was the most scrupulously clean and neat apartment I
ever saw, its pink window curtains casting a modest and
retiring light on everything, but, at the same time, beautifully setting off the snowy whiteness of the beds.   In
St Louis the Roman Catholics are numerically powerful
and very rich, and nothing proves more the never-failing
desire of those religionists to push the interests of their
faith wheresoever they can obtain a footing, than the fact
of their French Indian settlement on the prairies, "Potto-
waddami," so pronounced, at which I rested one night on
my return from Fort Riley.

On the afternoon of the 14th I accompanied Dr Pope
to a semi-private exhibition of the working of the fire-
engine attached to the station of the Fire Alarm Telegraph in St Louis. Nothing could surpass this establishment.   To the engine were attached four splendid horses,
all driven by one man on the horse at wheel on the near
side.   One of these horses was as magnificent and showy
an animal as ever I saw in a gentleman's carrriage in
England.   Of course I expected that to show me the force
and power of the water thrown by the steam-engine from
the long hose we should proceed at least to some uninhabited prairie outside the town ; but not a bit of it. We
marched but a short way, and then took up a position

in the thronged streets adjacent to water, and commenced
our field-day. At first the water came sparingly, and the
passengers went by unsplashed, and the boys and Boh-
hoys leaped playfully here and there over running but
diminutive gutters. At last the power of the engine
made itself felt, and, to my intense amusement, thundered
a volume of water on every adjacent window, carried off
tiles from the tops of private houses, shook chimneys, and
knocked away coping-stones; and then so terrified mules
and horses, and washed people who wished to exercise
their freedom or right of way in the midst of our aqueous
liberty and revel, that it rendered the streets around us
untenable to every living thing but ourselves, and at once
proved the indelible right of free citizens to do as they
liked without the slightest reference to others. Everything, however, went off in the greatest good humour.
Proprietors of houses put up their shutters, or shut their
windows and doors, till we had done, and the whole host
of Boh-hoys who collected to stare at us and frisk about
amidst the puddles and streams occasioned by our hy-
draulic experiments, however low their state and station,
could no more be taunted with the appellation of "the
great unwashed." Some account of the Fire Alarm Telegraph, with which this fire-and-water engine was associated, may neither be uninteresting to my readers, nor
useless in guiding us to as good an arrangement in our
cities and towns.

The fire telegraph in St Louis has, I think, been established for three or four years. It consists in a system of
telegraph wires connecting different parts of the city with
the main station in the north wing of the court-house.
There are five districts, in each of which there are an
alarm bell and several stations. Each station has an iron

box, always kept locked, the key in the care of a responsible person in the immediate neighbourhood. On opening one of these boxes there is to be seen a " crank," which being turned instantly sounds the alarm at the main station, whence the intelligence is transmitted to each district in the city by ringing simultaneously the alarm bells. When a fire breaks out at any point, the nearest box is unlocked, and the " crank" turned fifteen or twenty times round. This occasions a clicking in the main station and puts the telegraph apparatus there in motion. The operator on guard then examines a strip of paper reeled off by the movement, and learns by an impression on it in which district and near which station the fire is. He then sounds the alarm bells in the city, in such guise as to inform the members of the fire department where their presence is needed. If the alarm bell strikes four time ssuccessively, the fire is in the fourth district, if twice, in the second, and so on. The firemen being thus apprised of the locality of the fire, put to their horses, ever kept in readiness, to the steam fire-engines, light the fires under the boilers, and hasten to the scene of action. The intelligent and most capable superintendent of the telegraph, Mr James M. Gardiner, has a salary, I believe, of a thousand dollars. The alarm bell for the first district is at the St Louis engine-house, that for the second and third districts is the cathedral bell, that for the fourth district is the bell of St Francis Xavier, and for the fifth district the bell of the Mound engine-house. There are, if I remember rightly, about fifty station boxes in the city capable of giving alarms.

The time at last came for me to leave the city of St Louis, wherein I had made so many agreeable acquaintances, and wherein I had so much reason to admire the beauty and grace of the fairer sex, as well as the generosity, urbanity, and good fellowship of the men. Not in the least liking the lines of rails, their method or management, by which I had come out, with the kind advice of the Messrs Campbell, having expressed my servant and dogs right through to New York, there to await my arrival, I resolved on going the Canadian route, and thus to return by the giants of the giant land—the Falls of Niagara.

Before entirely taking leave of the State of Missouri, and in proof of the wondrous secrets yet to be revealed through the research of man in the womb of the past world, it may be interesting to make a brief quotation from the paper on mastodon remains found in the State of Missouri, and presented to the Academy of Science in St Louis.

Dr. Albert C. Kock stated, as published in " The Transactions of the Society in 1857," that " some twenty years before, he commenced making extensive researches and excavations for mastodon remains in the State of Missouri;" and the result of those researches were "some very striking evidences of the existence of man on the continent of America, in the age of the living mastodon. . . . In 1859 he discovered and disinterred, in lat. 38° 20' N., in the bottom of the Bourbeuse river, where there was a spring, the remains of the above-named animal. The greater portion of these bones had been more or less burned by fire. The fire had extended but a few feet beyond the space occupied by the animal before its destruction; and there was more than sufficient evidence on the spot that the fire had not been accidental, but, on the contrary, that it had been kindled by human agency, and, according to all appearance, with the design of killing the huge creature, which had been found mired in the mud, and in an entirely helpless condition.

"This," he asserts, "was sufficiently proven by the situation in which he found as well there those parts of the bones which had been untouched by the fire as those which were more or less injured by it, or in part consumed; for he found the fore and hind legs of the animal in a perpendicular position in the clay, with the toes attached to the feet, in just the same manner in which they were at the moment when life departed from the body. He took particular care, in uncovering these bones, to ascertain their position beyond any doubt, before he removed any part of them; and it appeared, during the whole excavation, fully evident, that at the time when the animal in question found its untimely end, the ground in which it had been mired must have been in a plastic condition, being now a greyish coloured clay. All the bones which had not been burned by the fire had kept their original position, standing upright and apparently quite undisturbed in the clay; whereas those portions which had been exposed above the surface had been partially consumed by the fire, and the surface of the clay was covered, as far as the fire had extended, by a layer of wood ashes, mingled with larger or smaller pieces of charred wood and burnt bones, together with bones belonging to the spine, ribs, and other parts of the body, which had been more or less injured by the fire. The fire appeared to have been most destructive around the head of the animal. Some small remains of the head were left unconsumed, but enough to show that they belonged to the mastodon. There were also found mingled with these ashes and bones, and partly protruding out of them, a large number of broken pieces of rock, which had evidently been carried thither from the shore of the Bourbeuse river, to be hurled at the animal by his destroyers; for the above-men-

tioned layer of clay was entirely void of even the smallest pebbles, whereas, on going to the river, he found the
stratum of clay cropping out at the bank, and resting on
a layer of shelving rocks of the same kind as the fragments, from which place it was evident they had been
carried to the scene of action. The layer of ashes, &c.,
varied from two to six inches, from which it may be in-
ferred that the fire had been kept up for some time. It
seemed that the burning of the victim, and the hurling of
rocks at it, had not satisfied its destroyers, for he found
also among the ashes, bones, and rocks, several arrow-
heads, a stone spear-head, and some stone axes, which
were taken out in the presence of a number of witnesses.
This layer of ashes, &c., was covered by strata of alluvial deposits, consisting of clay, sand, and soil, from eight
to nine feet thick, forming the bottom of the Bourbeuse,
in general, and on the surface, near the centre of the spot
on which the animal had perished, was situated the spring,
the water of which was used for domestic purposes; and
it was in digging to clear out the spring that the existence
of bones there had been first discovered by the owner of
the land."

Now, clear and fairly told as this account of the remains
of the mastodon is, combined with the proofs suggested of
the concurrent existence of that gigantic creature with
man, it is, nevertheless, so at variance with what has been
generally believed in regard to such co-existence, that,
without in any way doubting the truth of the belief of Dr
Albert C. Kock in the inferences he has derived, I am led
into the following queries.

We know that in America and in the grand features of
her colossal extent, the sites of rivers, mountains, and rocks
have changed, and that she, in common with all the rest

of the visible world, has been subject to vast eruptive and chaotic influences. That which is now a valley, through which I travelled in my way from Counsel Grove to Fort Riley, was once a river, as its alluvial bottom and rocks on either side the little valley prove ; and, from the existing state of storms and pervious nature of the alluvial mould pervading a vast extent of the plains, rivers may be said to have come and gone almost within the lives of men. That currents of rivers shift in the most uncertain methods and leave their beds to make new ones, I had a full opportunity of seeing from the deck of the steamers traversing that most fickle and dangerous of all streams, the Missouri river. Dr Kock says that the fire was chiefly beneath the head of the animal, and that the head was nearly consumed. Supposing the creature to have been stuck in the mud, the head, which might have rested on dry land, as well as from its position, would have been the only portion of the body beneath which a fire could have been placed ; but then the question occurs, was the ground dry enough, a swamp as it must have been, to permit a wood fire to have been lighted at the same time that the vast creature was alive, and, by the mere moving of the head from side to side capable of disturbing sticks or wood, put in a position to catch fire through the rude attempts at ignition within the power of savages ? Another doubt in my mind also exists, and that is, with the stated amount of wood ashes, would it have been within the power of such a fire so to have consumed the immense mass of green flesh and bones put but for a short time at its disposal ? There is again another position in which the two facts of the mastodon and the fire may be viewed. The bones of the mastodon only might have been there when the fire was lighted, and the Indians used them for fuel or as blocks through which to create a draught to nurse a flame, precisely as in this day we use either blocks of wood or stones. The stones or pieces of rock found among the bones might have been brought there for the same purpose. The heads of arrows, spears, and axes are things of such common occurrence that they only afford matter for favourable suspicion, or even but a straw to catch at towards the view taken by Dr Kock.

Dr Kock continues : " It was about one year after the excavation previously alluded to, that he found in Benton county, Missouri, in the bottom of the ' Pomme de Terre river,' about ten miles above its junction with the ' Osage,' several stone arrow-heads mingled with the bones of the same nearly entire skeleton (the mastodon), mentioned above as the ' Missouriam.' The two arrowheads found with these bones were in such a position as to furnish evidence still more conclusive, perhaps, than in the other case, of their being of equal, if not older, date than the bones themselves ; for besides that they were found in a layer of vegetable mould which was covered by twenty feet in thickness of alternate layers of sand, clay, and gravel, one of the arrow-heads lay underneath the thigh-bone of the skeleton, the bone actually resting in contact upon it, so that it could not have been brought thither after the deposit of the bone," a fact which he was careful thoroughly to investigate. " This layer of vegetable mould was some five or six feet thick, and the arrow-heads and bones were found, not upon its surface, but deeply buried in it, together with fragments of wood, and roots and logs and cones of cypress, but no pebbles were observed in it. Above this layer of mould there were six distinct undisturbed layers of clay, sand, and gravel, viz. three of greyish clay and three of pebbly gravel mixed with coarse sand, in all twenty feet in thickness, and a forest of old trees was standing on the surface soil.   This bottom is still subject to occasional overflow in very high stages of water."

It is this fact, then, the "recurrence of overflow in high stages of water," which in my opinion gives a strong ground to scepticism as to whether or not man and the mastodon had been in conflict and coeval in life the one with the other.   If the "bottom" where these remains were found is still subject to passing inundations, each leaving a deposit of one sort or the other, it must have been so at consecutive times, when the skeleton was much nearer the surface, though still enveloped in the clay in which it first became imbedded, and by the force and undermining nature of the water an arrow-head might have been washed beneath the heaviest portions of the animal structure, and there have remained to the present day.   The curious information given by Dr Kock is perhaps the most interesting that has yet re-awakened the minds of those bent on antediluvian research, and it may well give new life to a preadamite doctrine, as well as to the favourite argument of those religionists who pin their faith on the one great flood proclaimed in Scripture.   Nevertheless we must have further information yet, and fossil evidence too, of the existence of man in those remote periods, for while the bones of the water-rat have been fossilised, I believe the most perishable bone of all, known to be infinitely less durable than those of the human race, and not a vestige of man nor his cousin the monkey can be found similarly preserved,* why, great doubt as to the co-existence

---

* It has been asserted that fossil remains of monkeys have been discovered, but I doubt it.

of man and the mastodon will still cumber the assumed
history of the lost age. It is urged in worldly argument, attempting to strengthen that which ought to
need no crutch, that man may have existed only on
those portions of the globe now covered by the sea, and
hence his non-appearance as a fossil; but there are irre-
fragable proofs, and a voice in the stones, even on the
walls of the Cotswold Hills, as well as on the Downs of
Ashdown Park, and on the highest mountains, that the
world has owned a conquest by the sea, and to this hour
the handwriting of an overwhelming ocean still clings to
the topmost stone, and gives a graven autograph in
every little shell.

CHAPTER XXIII.

DURING my travels on the prairies, short as they of necessity were, I endeavoured to elicit from every man I met "notions," as the Americans would call them, of all he had ever seen or met with, always leaving a monstrous margin for the marvellous, in which it seems to be the delight of most of the inhabitants of the United States to indulge. In this I allude to certain classes of society among whom this curious recreation principally exists. The soldier and the gentleman lies not at all. It appeared to me that I could get just as much correct

information from a notorious liar off his guard, and
speaking sooth by accident, as I could from one of
greater general veracity on his purposed fashion of untruth, and I now quote from an unmitigated vagabond
and prevaricator, the man who was so improperly recommended to me as my guide, who really knew no
more of the prairies than pertained to the beaten path
which led to that bait or lure for industrious as well as
idle sinners — the gold-decked "Peak of Pike." This
man one day in conversation, when he had nothing to
get by lying, told me that he had a great friend of his
once living in the Rocky Mountains, who hunted a good
deal, and for many years "on his own hook," who had
told him of a vast deal of curious things to be met with
in that wild region. He never would tell me why this
friend of his was there, but I suspect that he had fled
from civilisation in haste, on account of the commission
of murder or other heinous crimes; and when I wanted
to know more of his history or his present whereabouts,
Mr Canterall persisted in it that he was dead.

Thus far, however, the information I sought for was
accorded :—This fugitive from his fellow-men reached
the mountains in possession of nothing more than a
blanket, tobacco, and his fire-arms, and used to live in
huts made of bushes, or sticks and grass, and in caves.
In his search for caves in which to shelter himself, and
which had always to be reconnoitred during the day for
fear of his becoming the sole lodger of a grizzly bear, he
came on a very large cave, more remarkable for the
things in it than the others. He described it as one "in
which the stones and rocks had been a great deal tumbled
about, and on the flat sides of some of them he saw the

huge footprints of animals he knew nothing of, and, on one occasion, *the largest print of the foot of man that he had ever seen in his life,* and that there were great bones there, such as he had never seen before." I asked my informant if he could in no way guess the direction of this cave, and told him that if he could guide me to it, on arriving at the spot, and finding that it really existed, and contained the things alluded to, I would not only go on to the Rocky Mountains, but reward him handsomely for having brought me to a site alleged to contain so many curiosities. To this he replied, " That his friend was dead, and had not described to him the exact position of the cave, and indeed, at the time when in conversation with the fugitive, he had taken very little heed of the matter."

Here then is intelligence which, if true, in all probability would connect the mastodon with man; and though I am no believer in giant traces of human beings antecedent to the one overwhelming flood, still, like a hare's foot on the soft snow, if a man trod on yielding clay, the impression made might assume an undue magnitude. Of this alone I am sure, that the vast territory of the United States holds on its surface, beneath its caves, and in its womb, not only a wider field, but a richer field for geological and antediluvian research than men are sufficiently aware of; and had I even now more available time and means for travel, nothing would interest me more than devoting some years to an attempt to reap, in the young country, the harvest of splendid and even awful research, the first seeds of which have been sown, though they have not yet ripened into perfection, in the old or mother country. By a further reference to " The Transactions of the Academy of Science, at St Louis," so kindly placed in my possession by Dr Pope, I find that the assumptions of Dr Kock, in regard to the coeval life of man and mastodon, are successfully disposed of, and that, however curious his research, and clever the inferences he derived from it, he has thrown no new light on the mighty mysteries of the dead world.

Having thus taken a hasty view of the wonders of buried worlds, as illustrated in the United States, or of many overwhelming floods, I now approach that, to me, most curious and striking fact, the existence in its present lightless world of The Eyeless Fish. That it has its world of waters to itself, or in conjunction, perhaps, with other blind things, there can be no sort of doubt; but how large or how small the world is—whither it tends, or what becomes of it—has not yet been revealed to the investigations of man. The stream where these eyeless fish are (for there is no mark of their ever having had an eye) rises in the pitchy darkness of the subterranean mammoth caves, and having run on the cavern floors for a certain distance, the waters, ere they see the light again, bury themselves in the bowels of the earth, and go we know not whither! I had not time at my disposal to pay these caves a visit, but I learned all I could in regard to them, and brought home a specimen which I obtained through the kindness of Dr Pope. To me there is a lesson in this eyeless fish which, in the most perfect way, and in regard to the meanest creature, proves the wise, all-knowing, and providential hand of the Creator, for here we have so legible a fact, that all who run may read, that an Omnipotent Being ruled that there *should live* a creature who, never aware of light, should not need the orbs to know the loss of the great luminary of day, but should for life remain utterly blind. The doctrine

that, from being in the dark, these fish should gradually
have lost their sight, and have blindly bred till custom,
assigned by place, should, through the force of inherent pro-
pagation, have induced no eyes at all, is a mere assumption,
and to my mind almost an endeavour to rob Nature of
her almighty care.   These fish are found only in that
dark world of waters to which the light never pene-
trates; at least it is not within my knowledge that
they have been discovered anywhere but in the
Mammoth Caves.   The fish, on a superficial view,
seems to me to be like our gold fish, or certainly
of that sort, but of course without their gaudy hues.
I believe the one brought home by me is equal to any
other ever brought to this country, if not the most
perfect of any; it is with the greatest pleasure that I
invite attention to so curious an investigation.

Adieu, then, to the good city of St Louis and to the
State of Missouri.

Two days' hard frost and snow had rarefied the air and
braced the nerves of men who had been suffering from
aguish fever, and on a Tuesday morning on or about the
14th or 15th of November, I found myself starting from
Barnum's Hotel, in an omnibus, a little before seven in
the morning.   Having taken a breakage-ticket right
through from St Louis to New York, for which I paid
thirty-one dollars, I boarded the steamboat and proceeded
about twenty miles by water up the Missouri, and as we
left the city had a very picturesque view of it from the
deck.   On the entire voyage I saw plenty of ducks and
other small wild fowl, and had a very pretty view, as we
approached it, of Alton, on the right bank of the river—
a high bluff promontory, jetting boldly out above the
stream.   Alton is a considerable town, but, like all the

towns of the West which I saw, too much scattered, and built without any visible design save to shelter heads as they chanced to locate on the spot. We disembarked at Alton about half-past ten a.m., and then ran on rail, chiefly through dead flats, to Chicago, 257 miles. On the journey I saw a common rabbit, such as we have in England, by the side of the rail. The line of rail I was then on was the best I had seen, and at last we travelled between sufficient fences. Nevertheless, while sitting in the extreme end of the carriage, I saw a horse lying in the ditch by the side of the rails dead, with his shoulder nearly cut in two, and covered with fresh blood. On halting at a small village or station to dine, I told the conductor, of the recently killed horse, when he accused me of poking fun at him, and swore I had been deceived. A very good chicken pie at dinner, but, as usual on the railway line, nothing to drink but tea and coffee.

Springfield Town, which we also stopped at, was remarkable for the neatness of its houses. Llangollen also was a neat and rising town; soon after passing which, and particularly at the decline of day, at every stoppage I thought there was a vast deal of needless delay. At half-past twelve at night we arrived at Chicago. Here I caught an uncomfortable glimpse of Lake Michigan, and then hurried to bed at the hotel, having to start again at three the next morning.

While on the rail waiting for a start, a poor but respectable Irish woman came into the carriage where I was, with four children, and attended by two Irish men, who came to take leave of her. While there, a railway official looked at her tickets, and said, that they were " second class." On this I went up to him making the remark, " I have paid a higher price, so show me, then, into a first-class carriage. It is evidently cheating if you take different prices for the same carriage." The fellow stared at me, and said he would see about it. However, nothing was done, and I took the trouble to ascertain that though they really took two 'prices, and regulated their charge according to the appearance of the traveller, there was but one carriage for us all. For some time we sped along by the shores of the splendid lake, and through a good deal of swamp and wildfowl-likely land. We then passed several tributary streams and rivers, promising trout and other fishing; mills upon some of them, and the water clear and brisk.

Having arrived at Detroit, I there found a capital hotel, kept by an excellent landlord and gentleman (Mr Fellers), called the " Michigan Exchange," to which I most heartily recommend all future travellers. Having left Detroit at eight a.m., we crossed the Detroit river in a steamer, and then, oh! what a comfort there was in recognising English railway uniforms, and to find myself on the Great Western line, and in Her Majesty's dominions. No sooner were we in Canada than a difference was at once perceptible in the face of the landscape. Fields were better laid out and fenced, all things wore a more cheerful appearance, and as to sheep, the superiority of breed was at once observable. The next piece of water to Lake Michigan, along the margin of which we sped, was Lake St Clare. The towns of Chatham, Kingsville, and Newbury were soon passed, and then we ran between woods and clearings in a sandy soil, on which the fir and other forest trees grow very well. To-day, at a little before noon, two ladies in black came into the train, the youngest of them so fair and strictly beautiful that I very nearly commenced a poem in her

praise. Hurrah for Canada! We then came to London,
a very large, well built, and excellent town. At twelve
on this day there was dinner for those who could eat at
that unfashionable hour, and on the platform there lay
three deer. They consisted of an old buck, of course utterly out of season, a younger one, about three years old,
very poor, and a young doe, the doe bad enough, but still
the best venison of the three. At a little distance from the
station I saw a very English-looking cemetery. While
stopping here I telegraphed to the Monteagle Hotel at
Niagara, to say I should want a dinner and a bed.
When we again started we passed the towns and stations
of Woodstock, Paris, and Dundas, the latter built in a
hole. Sir A. M'Nab's house in sight, at which I
should have liked to have made a call, then the very
beautiful Lake Ontario, which in scenery reminded me
of the lakes in Scotland. Thence I saw some stupendous railway works, and after that St Catherine's and
the Wellington canal. It was soon after this that I
beheld that most superb specimen of engineering skill,
the double suspension bridge over the Niagara river
below the Falls, the great bugbear of timid passengers.
This is a wonderful bridge — a lower passage for carriages, horses, and foot people; above the heavily thundering train; while beneath roars such a volume of
raging water that, what with the noise above and below,
Babel in comparison was the impersonification of silence.
Having arrived at the Monteagle Hotel, I found a late but
sufficient dinner, compared with the customs of the country,
or a supper which I made my dinner. The steward in
attendance was accommodating, and the girls who waited
at table nice and intelligent. Having ordered a fly for
nine o'clock the next morning, in order to have a whole

day for the grand aqueous sight of this part of the world, I found two comfortable rooms prepared for me, and all that I could desire, when, with a dreamy and distant but still defined sound of the Fall in my ears, I went to sleep full of expectation, and awoke in no one way to be disappointed!

By nine o'clock on Friday, the 18th of November, after a very good breakfast, I entered the comfortable pair-horse fly which, with an intelligent driver, mine host of the Monteagle Hotel had provided for me, and with a longing expectation I bade the driver take me to the English side of the Falls. On reaching the suspension bridge before described, a heavy train was propelled across the bridge above, as I came the reverse way in my carriage, while beneath us fretted, leaped, foamed, rushed, and out-roared the stupendous rattling thunder of the train above, the pent-up fury of the Niagara river. Nature's voice and artificial sound thus came in contact in an antagonism seldom witnessed, when thus early in the day it was given me for the hundredth time to hear as well as see how small are the ways of those that are bad when compared to the things that have been " seen to be good" under heaven. After passing the suspension bridge we ascended to the left, when, on coming to the summit of the rocky cliffs, 160 feet sheer above the boiling river, in a voice of admiration I called to the driver to let me out and hold his tongue.

In one of those still, soft, and lovely days, when mere existence is a blessing, when the air is redolent of health, and the heart of man disposed to feel the best influence of Nature, I stood above the Niagara river, and advanced so closely to the edge of the precipice that my well-intentioned coachman dared to disobey my orders, and *to speak* a word of caution. On the opposite side of the river

the banks and rocks were fringed with trees, while, on
the side on which I stood, bushes and fallen trees went
down to the water's edge, a few large trees still growing
on the cliffs at my side, on one of which I put my hand
to dispel any giddy sensation. While thus I stood, a
roar, still at some distance, told me that I only saw the
path to the lion's den; the lion had yet to be approached
and to be seen, to be thoroughly appreciated. While
looking thus upon the river, and at the attributes of its
beauty on either side, then on the American bounds
opposite, I had to lament the occasional erection of a
visible house, while close to me the trees which had
been so ornamental were ruthlessly cut down to afford
any fool who had built a cottage or villa a selfish peep at
the prospect which ought to be protected as the property
of the whole world.

Having admired the majesty of nature, and abused
the acts of men, to my heart's content, I again advanced,
bidding my carriage to follow, for it seemed, under all
the interest I felt, an act of sacrilegious folly to rush
rudely on such a scene, and when I stopped again I
paused on the spot where Blondin, a Frenchman,
crossed above the raging river below the Falls on a rope.
The flyman having risked his life again by breaking
silence to tell me this for a moment, I wrenched my
mind from the face of nature to the acts of men, and
asked him if he was present when the rash act was done.
He said he was, and moreover he pointed to an opposite
rock whence, he said, another man named Shields
had arranged to dive for public amusement, but having
come one day to take an initiatory leap into the water
in the way of ascertaining how much he could do, he
leaped in, disappeared, and continued to dive for three

weeks, turning up again only, or rather being found turning round in the whirlpool, a mile and a half below the spot of his intended exhibition, and then so disfigured that his identity was only guessed.

I did not wonder at the fate of this man after I had seen the Falls, and the river above and below them, for from the ocean of water, and the pent-up and rocky channel that receives it, its varied juttings out and deep recesses, the river below can only consist of whirlpools and conflicting waves, with such a tide above and under-tow beneath, as would overpower a million hippopotami. It is said that in that part of the river called, *par excellence*, " *the* whirlpool," there is from one to two hundred feet of water. Here in the river, hugging the American side, a small steamer plies, and disfigures the wildness and what ought to be the unaided loveliness of nature. As I thus advanced the roar of the Falls increased, and then they came so fully and so immensely, so grandly and so far beyond all powers of expectation, full into my astonished view, that I stopped, and, at least in heart, soul, and mind, endeavoured to take in or embrace the prospect. Stricken as I was with awe and wonder, and incapable of sufficient admiration and homage, a painful and a pitiful sense of shame crept over me, occasioned by my being forced to observe that, in the very midst of heavenly and beautiful things,—yea, almost in the centre of the wild and majestic Falls, where the waters leapt to an awful sacrifice of all life that came upon or within them, from the smallest fish to man, beast, and bird—the Americans without a head or heart, as it seemed to me, even here had " gone ahead," and by the aid of a few fragments of rock that yet looked up from the rushing waves, had from Goat Island crawled out—

for such progression can have no better name—and stuck
a sort of pepper-box-like tower right on the brow
of the cataract, whence they might smoke their cigars
and spit into the water. It only required a weather-
cock on the pepper-box tower to have stamped the whole
thing with a refinement of Cockneyism immeasurable !

The little steam-boat before mentioned, " The Maid
of the Mist," steams up as far as she dares go to the cliff
on the river approaching the Falls; and there again
American ingenuity and fallacious taste have run up the
face of the rock a little wooden staircase and mimic railway, worked by water—I should think by the tears of
the offended Niagara—to land unworthy visitors ! What
a teapot contrast that little steamer affords to the boiling
of this world of waters ! " Blotches and blanes," indeed,
have, on the American side of the wonder of the universe,
defaced the grandest natural feature man ever saw. I
wondered, and passed on, regarding the opposite side of
this picture as an American fall in more senses than one.
On the English side the worshippers of Nature are content
to gaze from the platform rock, or to stand beneath the
mighty Fall itself; they have defaced nothing. But if this
record of a flying visit reaches the eye of any member of
the Canadian Government, oh ! let the author pray that
protection yet may be afforded to the trees overhanging
the precipice above the river, coming from the suspension
bridge, and before the Falls are attained, for at present
they are, and are still being, defaced or cut down, to
please an idle wantonness or give a view to a cottage-
window.

From the Platform Rock I went immediately to the brink
of the monstrous cascade, and stood and sat and listened
to the voice of the waters, and walked and lingered and

loved existence, in that it simply gave me health to value all blessings and undivided appreciation of the splendour that was thus offered to my view. When I say " undivided appreciation," I do not mean to assert that I did not wish for other eyes that were far away to enjoy that prospect with me, for I did, and had they been there the world could have shown me nothing more. While looking on this mighty prospect on that still, sunny, and lovely day, there was around me, close above the Falls, a thick mist and a gently-descending rain. The rain is called the rain of the Falls, and as my soul drank in all that was offered me by Nature, for an instant reflection carried me back to my published opinions as to the fact that the mists which were the exhalations of the water in and on the earth ascended to the skies to be condensed and given back in rain, and that to the same extent as man drained the earth and decreased the exhalations, so would fail the amount of moisture rendered back by the skies. Around me ascended nothing more than a mist, over my head was a cloudless blue sky ; but when the mist reached a certain height, down it came condensed in very palpable drops—indeed, a very sufficient shower.

While occupied thus, a tall black man accosted me, asking "if I liked to venture beneath the Falls ?" when I replied, "Certainly, but how about the falling rocks above? My friends tell me that of late it is not safe to do so." I said this just to hear his remarks upon it, my friend, Mr H. Campbell, having strongly urged upon me not to go under them, for that a piece of the table rock over which the waters leaped had already gone down, and it was apprehended that more would follow. I had made no promise to Mr Campbell, for I was bent on seeing all I could. The black then told me that a piece of rock had

gone down, and made a division in the Fall above, which
he would show me if I made up my mind to go under the
cascade. "Go on, then," I said, "show the way."
"Stay, sir," replied my sable guide, "you can't go under
as you are; you must come to the hotel, and put on a
waterproof dress. On this I accompanied him to the
Clifton Hotel hard by, which is so near the Falls, that of
all places in the world at which I would advise deaf and
dumb people to sojourn the Clifton Hotel is the very
place, for cars and speech are almost useless there, while
signs and converse on the fingers are much to be coveted;
and in the black's robing-room I put on, from head to foot,
an entire suit of what once had been a waterproof dress,
and thus attired followed my guide down some steep steps,
and a passage made to the very foot of the mighty Falls.

It is impossible to describe that strange scene. At first
the noise is intensely surprising. It is not a splashing of
water—no waves in their angry ricochet are heard, and
no boiling or rushing currents; but there is one ceaseless
heavy leaden fall and roar, which drowns every sound
besides, and renders a voice, though speaking close to
the ear, almost inaudible. There was a yard or two of
slippery rock, enveloped in the wettest fog I ever encountered (to which a Scotch mist, though it wets a man
to the skin in a moment, in comparison is nothing), to
walk on; and by it I passed under the small portion of
the cascade which had been severed from the main body
by the fall of the portion of the table-rock alluded to;
thence I went on, followed, not led, by my sable guide,
till he touched me on the arm and made a sign for me to
stop. I did so, and contemplated the scene! At our
backs was the high, wet, black rock, and over-head
millions of tons of water; at our feet a slippery path of a

yard or so wide, and at its edge the boil of the sparkling, frothing tide that whirled up from its crushing fall on rocks worn fathoms on fathoms deep. Deep, deep, down fast and far below, and yet within a yard of us, fell that enormous avalanche of liquid death, for all of life that felt its force (could all life have a tongue) might well, with Byron, say, on encountering its rage—

Hope, withering, fled, and Mercy sighed farewell.

Having gloated on this strange sight, I turned from my guide to go on, when he touched my arm and roared something which seemed to be a caution not to go further, or, that that was as far as people went; but, on looking on the path and so understanding him, I saw it would yet afford space enough, with care, for a further advance; so, taking on myself the initiative, I proceeded some distance further, and had I not had nails in my shooting-boots, I could have gone a few feet further still. However, having nails, I let prudence be the better part of valour, stood still, and called my dingy, tall companion to my side. He came, and we stood shoulder to shoulder for some little time. In that short pause, and in the dim, strange place, I thought, as my guide was an athletic, not over well-looking, runaway slave, of the possibility, supposing I had been a less powerful man, of his seizing me, robbing me, and then pushing me but six inches on one side! How easy it would have been for him to have gone back and said I had fallen in by accident and been lost. There were no human eyes to see, no ears that could hear, and there would have been no signs of human violence on a body, if the body had ever been found, whatever signs there might have been of blows from the pointed rocks; and, thinking all this, I hugged myself in the knowledge that any attempt of the kind on me, if not met in time by a quick, straight blow, would assuredly entail a double death; so, deeming myself quite safe, and the thought of just enough value to enhance the beauty of my position, I remained a little longer, caught a small cray-fish on the rock at my foot, and then motioned my guide to return, as there was no passing each other there, nor any possibility of my taking precedence. When we got back from beneath the Falls my guide told me I had been further than usual, so I returned by myself and stepped the distance, which, without including the outer cascade, separated by the fall of a portion of the rock, was upwards of twenty-seven yards.

On taking off my said-to-be waterproof dress, I found myself quite wet through, when, having dismissed my guide, I again went above the Falls, and walked some way by the side of the river as it rushes to take the final leap. On returning I observed that, as a protection to the bank, just on a level with the Falls, one or two small sort of stone piers had been made, and by leaping from one to the other I attained the outermost, and stood on it, just above the Falls, and with the water rushing by me. Among the loose large stones of which this pier was made I found one with the complete ossification of the mullet on it even to the feelers—I suppose I must call them—at either corner of the mouth. With this treasured reminiscence of the place, though very heavy, I contrived to leap back, when by a whistle I procured the attention of my black friend, and desired him with a hammer to sever the fish from a large portion of the rock attached. My black, pleased, I suppose, with former liberality, hit the stone, in spite of my caution to be gentle, so zealously, that he split my fish, but not so much as to destroy its proportions, and I brought it, with my little cray-fish, home. By the splitting of the stone, however, I obtained a beautiful piece of dog's-tooth spar, which was very well worth having.

By this time hunger told me that it was time for luncheon, so I proceeded to the Clifton Hotel, and after lunch showed my stone mullet to mine host; he had a room of similar collections.

"Good gracious!" cried this veracious Boniface; "well, sir, you have been lucky! What would I not have given to have been the fortunate finder of such a treasure as that; good gracious! it is worth twenty pounds. But pooh, you would not take twice that sum for it."

In return for this beautiful "gammon," I quietly remarked, "I like the specimen much. *Is* it really in your *honest opinion* worth twenty pounds?"

"Worth twenty pounds, sir," he cried, in feigned excitement, and, calling to smartened-up barmaids to back him with their blandishments, "Look here, my dears, I should be glad to give this gentleman twenty pounds for this specimen; but, oh, no, I am sure he would not take it."

"There you are mistaken," I replied; "perhaps I can find another; and as my heart is so opened by the beautiful things I have seen, and I wish to do you a kindness, and make it even better than your own bargain, give me ten pounds and the specimen is yours."

I never saw a man look more foolish, as he hurried away, saying, "No, no, no. I'm sure you won't take it." "Yes, I will," I said; "if you will give it. Nay, more; so great is my wish to do a kind thing, you shall have it for five," when my offers were cut short by his plunging down a side-door or trap, where I thought it useless to follow him.

Adieu, then, to the English or Canadian side of the Falls! With difficulty I tore myself away, and, walking back, I bade my carriage follow till I came to a road in the side of a hill leading down to a ferry below the Falls, where the expanding cliffs afford more room to the rushing tide, and it is possible to row a boat across from shore to shore. Telling my driver to wait, I had some inclination to climb down the cliff the shortest way to the water, and for that purpose, my coachman staring at me in astonishment, I clambered over the bank that protected the road; but, observing that the face of the descent was very loose, and that I could not be in command of my legs, I clambered back again and went down the legitimate way. At the margin of the water I found a small ferry-house and the ferryman, with whom I entered into conversation, and the first question I asked was, "Why did so many seagulls of the common sort keep perpetually hovering on the edge of the fall of the waters below the cascade?" "Don't know, sir," was the intelligent reply; "they always do from the fall of the year till towards the spring." "Humph!" I said to myself, "but I guess why. Do you ever find any large fish killed by the Falls, and washed up here to your own door by this strong eddy?" "Oh, yes, sir," he rejoined, "it is quite worth my while, as well as that of others, to examine the shores and eddies hereabouts every morning, for we often pick up large fish and all sorts of wild fowl, and tame geese and ducks that have been killed in coming over." "What sort of wild fowl?" I asked, thinking it strange if wings could not save from this destruction. "Oh, many sorts, sir, wild duck, divers, and others; there were eleven tame geese from the farms above killed in one night, and once

26

as many as thirty wild ducks." "Oh, oh," I thought, "I must muse on this;" so picking my way over slippery rocks and leaping from one to the other, I got out as far as I could into the river and sat down for further contemplation. My first occupation was to observe the gulls, and from the spot to which I had thus attained, I could see them watching on the boiling-up tide at the immediate foot of the cascade, and picking up little fish killed by the weight and force of the Fall, so that the why and wherefore of the presence of the gulls on the spot was easily settled, their temporary absence also accounted for by the necessities of the breeding season. But how to account for the death of wild ducks—I could understand the destruction of tame geese, and for the deaths of the great Northern diver, which I subsequently ascertained so frequently occurred—that afforded matter for consideration. I could not solve the point then, so must recur to it hereafter.

Having spent some time here, I ascended to my carriage, and ordered the driver to stop in passing at the house of that clever naturalist and bird-stuffer, Mr James Booth, residing at the Falls, Canada West. Here I spent some time in looking at his valuable collection, and at a specimen of the largest beaver I ever saw. From him I procured two or three specimens of . birds in better plumage than my own of the same sort ; and he put the little crayfish I had caught into a bottle of spirits of wine with a young cow-snake, picked up in the road leading to the ferry, and to those added a beautiful limestone impression of the fossil sycamore leaf. I left him much pleased with my visit. On recrossing the suspension bridge, my carriage pulled up on the American side, at what I suppose is a species of custom-house, for the driver in-

formed me that I should be searched. The officer, however, was very civil, and taking a slight glance at my
effects, was satisfied, and I passed on. In crossing the
bridge to "Goat Island," which is an American possession,
and which divides the Falls, a country and county man
accosted me by the name of Hicks, whom in former times
and at elections in the western division of Gloucestershire
I well remembered, and I gave him all the news I could.
Having left my carriage, I was also spoken to by a man
who was once a farmer near Cheltenham, but now a fly-
driver at the Falls, and of course I gave him a trifle,
the receipt of which was not unbecoming his present state
in life.

A more lovely spot than Goat Island could not possibly
be, but alas, oh, for American taste! right in the middle
of the otherwise picturesque river, on a rock washed on
either side by the stream as it rushed by, Yankeeism had
once more gone ahead and perverted a portion of the
stream to coin its dollars through a mill! "Oh for a
cannon," I cried, "to knock that citizen up to the top of
some high hill, where the wind might play on him instead
of the water for him!" I now inspected the magnificent
American Falls from the foot of the island, and stood on
the spot where an indiscreet man had swung a child, by
way of amusing it, over the terrible cataract, and then let
it jump out of his arms to be dashed to pieces. From the
foot of the island I repaired to its head, and walked out
and ascended the American pepper-box-like tower before
alluded to, much like a nasty pimple on a man's nose, to
render as hideous and unnatural as ingenuity could make
them the rest of the features. While up in this tower I felt
ashamed of myself in having sought the view by such a
means, and hastened back again. By the view from this

tower up the river before it reaches the Fall, as well as from the higher end of the island, however, I came to a clearer insight of the volume of waters so precipitated and in such immeasurable quantity. The view of the river above the English Falls presents the appearance of a raging sea, or of a " ground race," which, in fact, it is, and it is a succession of continuous small falls, all heaving the dashing waters towards one grand climax. I can then easily conceive that wild-fowl imperceptibly hurried on, and, confused by increasing distractions caused by the leaping up of the mad waters and the general roar, permit themselves to be carried downwards, when, as the crest or brow of the enormous wave, just as it curls to fall, offers a smooth space, the fowl, not aware of the imminent ruin that lies below, and that they do not then see, get into the terrible impetus far enough to be suddenly aware of their danger, and turning against stream, and stretching their wings, the mere splash of the surface spray strikes them and hurls them to perdition.

About seven years previous to my visit to the Falls, and just at day-break, or about five o'clock in the morning, a man was proceeding across the bridge on the American side to Goat Island, when his attention was attracted to an unusual object rising from a rock, scarce five yards in circumference, in the very midst of the raging river, and scarce two hundred yards from the brink of the awful Fall. The man stopped and stared at the dark object, for it had the outline of the human form, though the situation where the figure appeared was supposed to be beyond all living or human reach. He stopped and stared, and, staring, doubted his senses; but as every instant the dawn of day increased, he felt assured, in the increasing light, that the object which he saw on that ex-

traordinary and slippery foothold was a man, and that
that man saw him, and, though benumbed and in a feeble
state, half mad he waved and wrung his hands en-
treating for assistance.

To rush back to the village and alarm the inhabitants
was but the work of a few moments, and when the entire
population arrived upon the bridge the morn was brighter, and with a thrill of horror they beheld a man upon
that little rock, each welling-up of the whirlpools laving
his very feet, and resembling chilling messengers from
the angry roar below, sent to claim him for destruction.
There was no mistaking the signs for assistance continu-
ally made by the lonely, lorn, and wretched figure; his
voice was already drowned by the shrieking waters, when
a spontaneous cry arose from the lookers-on for a reward
to any one who would save the miserable being. A sum
was named as the reward of rescue, and an American—it
is just to humanity that I should name his country—
volunteered to save him for that sum; and in a short time
produced a raft and a length of rope, which, let down
from the bridge, could be guided to the miserable sufferer.
There was, however, one present who had saved more
lives in that river, and knew the currents, whirlpools, and
strength of the tide better than anybody else. This was
a Mr Robinson, a carpenter by trade, and he warned the
people that the only way to save the man's life was to drop
the raft, after it had reached the rock and received its burthen, down to a little island nearer to the main island,
from which, at the risk of his own life, he engaged to take
the man in safety in his boat. But no; the wretch possessing the raft refused this wise and humane counsel, on
the plea that he had undertaken *the job,* and that no other
person should interfere, nor share with him in the reward.

There was no reasonable man there to put a revolver to this villain's head, and take the rescue from him; the crowd were, I suppose, horror-stricken, or by the mere force of numbers they might have attended to the good adviser of the boat, and lowered the raft accordingly. Guess the thrill of intense joy in the wretched man's heart when a chance of rescue came, and he heard and saw the raft knock against his slippery little rock. The overwhelming force of the tide he had already felt, as it struck him against the place to which he had succeeded in clinging, and thence, and up to the very instant of time of which I am writing, he had listened to the hoarse roar of the monster just below him, who, the greater part of the night, and all the morning, had seemed to be crying out for him *to come*. Knowing the force of the tide, he waved his miserable hands and almost fell from his little vantage-ground in the efforts he made to induce the owner of the raft to insure his life by dropping him to the island. He could not know that his suggestion was refused, he could not know that his signs and the prayers of the by-standers were all in vain; he stepped on to and clung with the tenacity of an ebbing life to the rope and raft, but in despair, for he found that he was hauled up towards the bridge. Vain were his convulsive grasps, and vain his shrieks; the edge of the raft, when pulled against the rushing stream, of course succumbed, and with a horrible flourish the poor wretch was whirled from his last hope of safety to his now certain fate. A strong swimmer, he manfully struck out, in the hope of slantwise reaching the shore. It was a terrible race to see, but soon it was all one way; the tide prevailed, and he *must* go over! Not till he reached that smooth and more than elephantine brow, where the volumed waves of the mighty flow curl

to take their last plunge, did he give up the struggle.
Then, when he saw his dreadful death in that smooth
space, he shot half his length up from the water, and,
raising his hands in despair high above his head, he took
the maddened whirl, which dashed him into a shapeless
mass, and left the sordid brute with the raft to all intents
a murderer.

All that could be gathered of this murdered man's history was that he, with two companions, had been drinking at some place on the previous evening, and that at
about dusk they all three left in a boat to cross the river
at the usual distance above the Falls. From that time to
the appearance of the victim on the rock nothing is known,
but three days after the catastrophe I have thus related from the lips of a living witness, three masses of
bruised and broken flesh were found below the Falls
near the suspension bridge. Whiskey and tobacco,
those inveterate destroyers of American men, no doubt
stupefied these people sufficiently for them to let their
boat drift till it came within that sweep of waters that
nothing can withstand, and then, either by the upsetting
of the boat on some of the minor falls or by their attempting to swim, two of them met their deaths. The most
wonderful thing on record is, how that one man managed
to lay hold of that rock, flat and slippery as it was; he
must have been flung against it, and perhaps wrenched
off every nail on his fingers in making a desperate grasp.

The little islands on the American side are lovely, and
fringed to the water's edge with cedar trees. Until I
stood at the higher end of Goat Island, amazed as I
had been, I had had no conception of the volume of water
hurled over the Falls, or of the breadth and rapidity
of the river where it separates to rush past the island.

Then and there the mind of man may compute the majesty of all he has seen and sees. Mr Robinson once took a man off in his boat from the middle isle of the Three Sisters, who had got there in some unaccountable manner; and on the 1st of September, previous to my arrival, two men in a boat were dashed over the Falls again, to a whiskey-and-watery grave. From the brink of Goat Island, immediately above the edge of the Falls, I brought a young cedar tree to England, but the journey was too long for it, as it was not in a pot, and it died. While examining the river at the foot of the Falls, I observed that there were some splendid eddies and backwaters, deep, and sleepily free from the boiling current, that promised fish, and I am sure that in the midst of this splendid scene, every sense astonished and amused by the magnificent panorama around, and refreshed by the pure air that seems to me to haunt that region, the angler, with spinning bait, might have some good sport with the American perch or bass.

I roamed in these enchanting scenes till dark, and then repaired to mine inn, unable to restrain myself from the repetition of the assertion that this "is a place for lovers and lovers only," and resolved to advise all lovers to try a honeymoon at the Niagara Falls.

CHAPTER XXIV.

HAD I had it in my power, I would have remained some
days at Niagara, but time and the Cunard steamer wait
for no man, so writing to decline all lectures, at Philadelphia and elsewhere, at five in the morning I took the
train for Rochester, saw, in passing, the Falls of Genessee,
and the river of that name, and then the station at
Lyons, then the Clyde station, and on by Onondago
Lake—or a lake of some such name—to the Salt Works,
and to Syracuse. The town of Rome and its station
were then passed, and we dined at Utica. Then came

the river Mohawk (I believe a branch of the Hudson),
and the Mohawk valley, Little Falls, town and station.

On this journey considerable crops were standing over
a large tract of country, the nature of which I did not
know; but on inquiry, I found that the Americans called
them "Broom corn," the stalks being manufactured or
tied up into brooms, and the grain partially used for
pigs, fowls, and cattle. On expressing my ignorance of
the crop my companions were surprised. "None on it
in the Old Country," they exclaimed; "Guess you're
at a loss then to clean your house?" An explanation was
necessary that though we had none of that sort of broom
in general use, yet still we had the besom and no end
of soap and scrubbing-brushes.

We continued our route by Amsterdam, originally
settled by the Dutch, and on till at Albany we got out of
the railway carriage, and crossed the Hudson in a steamer,
where we were again delivered to the rail. Albany is a
bustling town and port, and it was very easy to see that it
had a flourishing trade. We now coasted the mighty
river of the Hudson, and from having heard so much of
it, as to the Hudson only was I somewhat disappointed.
To my chagrin it now began to get dark; all the railway
acquaintances I had made had left at Albany, and I
found myself in a long carriage crammed full of all sorts
of passengers, all more or less spitting tobacco-juice;
three stoves in the carriage, one at each end, and one in
the middle, the two former being lighted, and serving as
a mark for all men to spit at, the hissing noise that tes-
tified a hit serving the purpose of the flag at a target
for rifle-shooting. The steam and effluvia thus arising
were so horribly offensive that I contrived to get a small
seat which only held two, close by the stove that was not

lighted, and an Irish emigrant, I suppose—a civil, good-humoured fellow, in the, I should say, labouring or farming class of life—sat by me. He shared some sandwiches I had, and offered me some whiskey, and our seat of course commanded, and had a right to command, the two windows next us, which were isolated and confined to us by the presence of the stove. It was nearly dark, and so hot, and the place so fetid, that we agreed to open our windows — they were close together. His would not open, but mine would, so I set it up as high as it would go. I then fell asleep, feeling weary as well as lonely — there is nothing more lonely than finding oneself in a crowd entirely strange— and how long I had thus dozed I know not, but I was suddenly awakened by a dreamy conclusion that an insolent man, with a displeasing flourish, had stepped across my legs, and, without a word, or saying by your leave, loudly closed my window.

The phase of dreamy conclusion at which I thus at once arrived was that, in my presence and person, all England had been insulted; so, springing briskly to my feet, with as manifest a flourish I flung my window up again, and then, resuming my seat, said with some emphasis, "No man in this carriage shall close that window—my window—without my permission." Perhaps I was wrong in this wide assertion, and it would have been more prudent to have confined myself exclusively to the one aggressor, and I became aware of the way in which my remark might be interpreted, by a wild sort of warwhoop from all the remote or dark seats of the carriage, but no whoop very near me. On this a nasal voice from the other side the carriage from the man who had closed my window, and who had no earthly

right to it, exclaimed, looking towards me, and in
reply to me, " And that shows the stuff you're made
on." " Very likely," I said, " but if you apply
the word ' stuff' invidiously, I throw it back, with all
the contempt it deserves. I say again, no man shall close
my window without my permission." Another warwhoop
from all the dark places of the carriage. " Wall! we
shall see," continued this snarling, quarrelsome voice;
" I 'll have that window shut down." A pause then occurred, in which the Irishman at my side pressed his hand on
my thigh—I suppose to keep me quiet. " Wall," continued
the cantankerous voice again, " we shall see when the
conductor comes if we don't get that window o' yourn
closed." No answer; I sat perfectly quiet, with it wide
open, and the snarling voice continued helplessly on
till I suppose it got tired, and then going on again, it
said, " Wall! we shall see if the conductor thinks that a
passenger has a right to keep his window open to the
annoyance of a lady." Then, for the first time, I perceived that there was a female at his side. The man then
remained quiet for perhaps ten minutes, when, seeing that
he had given it up as a bad job, I got up and crossed the
carriage to him—and a pin might have been heard to drop
when I did so—" Now, sir," I said to him, " we have both
had time to cool. If, in the first instance, you had civilly
asked me to close my window because it annoyed you, I
might, perhaps, have done so; but that time is past as far
as you are concerned. If, however, the lady at your side
even now tells me that she wishes the window closed I
shall be happy to comply with *her* request." I think he
said something to the female, and then remarked, " The
lady does wish the window closed." On this I strode
back and shut it, saying, " There, I have closed it for

a lady's sake, but no man in this carriage should have done so without my leave." On this the warwhoop again arose, but nothing further.

Soon after this little agreeable episode in American railway history had happened, I fell asleep again, from which slumber I was once more disagreeably aroused by the nasal tones of that wrangling and quarrelling voice. At first, of course, I thought that I was the object of its hostility, but on turning my head I perceived that the quarrelsome owner of that voice was struggling with a tall countryman of his as to which way a seat should or should not be turned. Each had hold of it, and each was trying to turn it the way he desired, and the struggle became most amusing. "I have a right to turn the seat, sir, and I will," said the tallest man, butting at it like a sheep. "Guess you won't though," snarled the quarrelsome man; "and mind, sir, take care what you are about, sir, or I reckon I'll get at something else, sir." "Will you, sir?" replied his opponent, "Guess you may get at it as soon as you please—guess I'm as ready as you." Then at it they went again, pushing like two bulls of Bashan, interlarding their struggles for the turn of the seat with many suggestive threats. At last I was so sick of the scene, that I was just about to get up and say, "Come, come, enough of this, let an Englishman settle the matter in a good-humoured way," when, to my intense disgust, one of them said to the other, "I'll bet you sixty dollars I get the seat." This allusion to dollars was too much for the national feeling, and on the mention of a wager up jumped half the passengers in the train, and began to bet upon the belligerents. I got up too to see the fun, and made one of the half circle around them. After one or two struggles more, however, the tall customer, who was perfectly in the right, as the quarrelsome man only claimed the second seat for his heels, gave in and retired to a seat hard by. The quarrelsome man, evidently apprehending a *ruse de guerre,* then sat on the seat they had struggled for, transferring his heels to the side of his female companion, in case his opponent should renew the contest from the other side. In order that he might do this effectively, he took care to sit with the back of his head right in front of the other man, so that his competitor should have a continuous view at least of the place where men's brains usually are. I kept looking at these fellows, till all at once the face of the discomfited hero lit up with a brilliant idea. The place which he had thus been forced to occupy commanded a window, which, when opened, would just let the wind and rain, then raging, on to the back of the annoying head. To be aware of this was to do it; the rain rushed in, the head reeled about in agony; rage was rife, but the elements prevailed; the water of course had its double effect, and displaced as well as cooled the angry man. All then remained quiet till just before reaching New York, when the conductor or superintendent of the baggage entered, and, coming up to me, said in a loud voice, " Your name, sir, if you please, and your address, that I may deliver your baggage correctly." Every soul within the long carriage was hushed to hear the reply, which was given fully and loudly for the benefit of all hearers.

When we came to our journey's end, and had got to our feet, the most civil consideration was extended to me, no jostling, no rudeness, but, everybody that I came near stepped on one side most civilly to let me pass, and I saw nothing more of the cantankerous fellow who it seemed to me, and greatly to the annoyance of his female companion, would have quarrelled with his own shadow.

Since this event I have heard from American gentlemen that these sorts of disgraceful scenes are frequent on the American rail, as, indeed, they must be when society is so improperly mixed up, and tobacco, to some exciting, and not of sedative properties, so fervently and so feverishly indulged in. I had hoped to have run through the United States without any "difficulty" arising between me and a decently-dressed man—Boh-hoys I count as nothing; but in this, the very last stage of railway travelling, I was doomed to be disappointed. In the morning previous to this row a window, set open by a man having a just command over it, did inconvenience me, and two kind and agreeable friends of mine then in the train asked me "if I would have it shut." I replied, No; I had no right to demand it, and, as I could not insist on compliance, I would not make the application. They, however, in their kindness to me, asked the man to shut the window, but they met with a flat refusal. All this tends to show the grave mistake in the American railway companies in not having a first, and second, and third-class carriage. I am told that, in their present ill-arranged state, but few of the rails are remunerative; why not, then, try a different plan? In England, in a first-class carriage holding six or eight people, there is never any dispute, the two windows being considered, as to their opening or shutting, as at the command of those sitting with their face to the engine, and where the current of air has the most power. If all sorts of people are promiscuously huddled together, in a carriage where there are a multitude

of windows, all more or less affecting those who do not
sit in the position supposed to command them, there must
be quarrels.   In my instance, however, a man from the
opposite side sought the difficulty—the error is in not
permitting gentlemen willing to pay to travel in
civilized society.

At a little after ten at night, a fly set me down at that
perfection of an American hotel, to which I advise all
travellers to repair, the Clarendon; and there I found my
friend and fellow-voyager, Mr Brown, waiting to accompany me back to England.   I also found that Mr Palmer,
the president of Sharp's Rifle Manufactory, had, with a
very handsome letter, presented to my acceptance the
newly-improved rifle carbine.   This carbine is not only
a breech-loader, but, if still more haste is required, on
turning one or two (rather too intricate) bolts or springs,
it contains caps, and will prime itself.   My opinion of
this part of the invention is, that the motions to make it
a self-primer or capper are far too minute and difficult to
be of any use on service, and utterly beyond the large
thumb of a trooper, particularly when anxiously under
fire.   Two other faults it has; and the first of these is
that the trigger pulls too hard, the second that there is
not sufficient room between the trigger-guard and trigger
for the first finger to lay all its stress on quite the end of
the trigger.   Other than this, it delivers its ball hard and
true; and I would call Mr Sharp's attention to a remedy
for these errors.

At New York my dogs and servant had safely arrived,
and with their larger boxes or kennels had taken up their
berth on board the " Asia," commanded by Capt. Lott,
where, on the 23rd of November, at 2 p.m., I went on
board, and shortly after bidding adieu to New York, where

on my return from the plains I had spent more than a happy hour or two, we were steaming down the harbour for the open sea.

In the last few days of my stay in New York I had paid a visit to Mr Galbraith, the naturalist and preserver of birds, in the Broadway, and from him also procured better skins than some of those I had myself obtained; for at the time that I shot them, most of the birds were on the moult, and worthless for preserving. Mr Galbraith was most attentive, and accorded me all the information in his power. I also paid a most interesting visit to the rare collection of creatures connected with natural history in the possession of Paul de Chaillee, and there saw several specimens of the gorilla, as well as an antelope, quite new to me. With Paul de Chaillee I had much conversation, and promised to try to get him an assured sum if he would bring his collection to England, he being too poor to risk his all in the matter; my endeavours however have not met with success. He told me that an old male gorilla had seized one of his men, a native, I believe, by the side with his short-fingered hand—the hand of the gorilla is shorter than that of the monkey tribe generally—and that with his hand alone he had broken through the outer skin of the body and torn out the man's entrails. Without doubting Paul de Chaillee's veracity, I think there is some mistake here. I deny the power of the gorilla to do so; and my opinion is, that a man's arm—a strong muscular man's arm—is more powerful than the arm of any gorilla. I think the gorilla must have used his teeth. I have often given my hand to the largest monkeys in menageries to test their power, but I had not the least difficulty in the retention of my finger from the bars of

their cage, to which they tried to pull it.   In New York,
at the office of Mr Wilkes, I met Heenan, and, without
Heenan at the time knowing who I was, we stood up
to each other in our attitudes of fistic defence.   As to
my opinion of his looks, his shape, and manners, it
has been previously and favourably expressed ; but
at the time I thought his attitude so open and faulty
that Sayers would have an easy conquest.   So faulty
did I deem it that I thought it was not his real position, but that he stood thus in order that an Englishman might not carry home any description of his tac-
tics.   By the photographs, however, that I have seen of
him, he holds his left hand still as low as he did with me,
and, what is more odd for an exclusively left-handed hit-
ter, which he is, he stands with his left leg and left hand
foremost.

But, adieu to New York and the United States!   We
are on the broad bosom of the Atlantic, with a nice little
company of voyagers and very agreeable and graceful
society.   During the voyage we had some roughish
weather, but not a gale of wind.   One rather stormy night
the lesser puffin-auk flew aboard of us, and was brought
to me.   I had him taken the greatest care of, and for a
few days he fed very well, but as sailors do not discrimin-
ate what portion of their grub is fit for feathered stomachs,
and what not, he ate something which disagreed with him
and died.   Again, one evening, long after dark, and while
the ladies were dancing on deck, a little female furzechat,
hundreds of miles from any coast, flew against the rigging
and dropped dead at a very pretty foot, from which en-
viable situation it was delivered to me: and these were
the only two ornithological incidents that came under my
observation.

Among the passengers on board the Asia there were
very few Americans, and indeed only one bad spitter,
who, with the usual contempt for all decency and the customs of other nations, continued to soil the smart floor.
Like his compatriots, he would not even condescend to
spit over the side of the ship into the sea.   One morning
after breakfast, on coming on deck, I found Capt. Lott
standing warming himself, with his back against the chimney, so I went and took a place at his side.   Presently after
I saw *the* American hat that covered the human syringe
of a head coming up the stairs, but ere it had ascended
much above the level of the deck it stopped and curiously
regarded, and I thought moved, to give way to something,
or some other than its own body, immediately behind it.
Having thus stopped in momentary observation, it began
its ascent again, and when its knees were on a level with
the deck, the body and head too turned round and again
contemplated some coming event whose presence was not
yet indicated by any shadow, and again the head and
body made way for whatever was behind it to pass, but
nothing came.   With a sort of desperation, then, the inveterate spitter stood on deck, and instantly there appeared the round, rough bullet-head of some sailor behind
him, on his hands and knees, and in his hand a " swab."
The spitter again curiously inspected the fore parts of this
figure, for it was only half on deck, but made no more
motion for room for it to pass, as there was plenty of room,
so the spitter walked the deck, but ere he had gone half
its length, again stopped to contemplate this very queer-
looking individual, who crept close after him like a dog.
The spitter then walked on at a brisk, after-breakfast pace,
but walk where he would, or as fast as he would, his shadow
on all-fours stuck to him like a leech, and on one or two

occasions on a cessation of speed, very nearly butted the calves of his legs with his head. The spitter grew desperate, and having forgotten, in the contemplation of his strange follower, for some time to relieve his laden cheeks in the usual way, he let fly such a quantity, that I heard it slap the deck as if a pint of water had been dashed upon it, and I think, his jaws thus cleared for action, he was about to question his pursuer, but there was no need, for, the instant he had thus soiled the hitherto beautifully clean deck, the amphibious sprite, on its hands and knees, sprang like a tiger on the offending stain, and removed it with the swab. This done, like a careful retriever he came to heel, and wherever our spitter went, again butted at his legs. The hat of the American had by this time become more limp than ever; having been angrily and frequently hit by the free hand of the free citizen, it assumed such shapes as only the tall chimney-shaped hats of a model republic can assume. The eyes beneath it became wild, and again was a triple and collected avalanche squirted under violent rejection on the decks, and again and again the embodied sprite of all tobacco sprung from the heel, obliterated all national stains, and again retired. I looked at this, I looked at Capt. Lott, I looked on the short, thick, bullet-headed individual who seemed to have devoted himself to transatlantic legs for life, but not a smile could I detect on the lips of captain or man—everything was gravity itself. The American was beyond the power of speech, and had the farce continued he must have thrown himself overboard. I could stand it no longer, and I burst into such an immoderate fit of laughter, that I turned my back to this strange struggle, and held the rail till it shook again. When I looked up, the time for ejection of juice had again arrived, and this time the

American walked to the bulwarks, and spit into the sea. When this was done I thought there was a slight caper from the man at his heels — a sort of jumping all-fours, as a dog will do when he expresses pleasure ; at all events the American looked round and down on his follower, and if the jolly tar had had a tail, no matter where, I am sure he would have wagged it. All at once the fact seemed to come across the free mind of the United States, and the lofty scion of those who despise labour seemed to comprehend that, to be rid of a man on all-fours at his heels, he must decently demean himself, and not disgust ladies and gentlemen. He again spit into the waves instead of on the deck, and then went below, his shadow following him, not on all-fours, but as a man proper and perfectly unconcerned.

" That is the best way I ever saw," I exclaimed, " to cure a man of soiling the decks," and again I roared with laughter.

During the voyage I was also charmed to hear Capt. Lott reprove men for not taking the trouble even to throw the ends of their cigars, still alight, off the deck into the sea. If people will indulge in smoking as well as in chewing tobacco, surely they ought to do all in their power to render their filthy habits as little dangerous and annoying to their neighbours as possible, and better order ought to be enforced in all ships where a medley lot of persons, perforce, are associated together.

The voyage home was prosperous enough in weather. There were a few briskish gales, but nothing of any consequence ; but during the passage I was attacked with so terrible an acute inflammation in both eyes, that for two days and nights I was blind, and in the most intense agony. From my fellow-passengers I received the kind-

est attention, and the steward was one of the best nurses
I ever saw ; so that, save and except the confined nature
of my berth, and the additional difficulty of keeping my
sea legs on in a sea way, when I could not see where I
was going, I had no great reason for lamentation : icebergs
and whales had been left behind, or had got out of season,
therefore on deck there was nothing new.   What I
lamented, of course, next to the loss of the ladies' society
on board, was not being able to see the first glimpse of
the coast of Ireland.   We landed, however, at Liverpool
at dark, in the first week of December—I think on the
5th of that month—my friends piloting me down the
gangway into the tender, when on reaching the quay I
got into a fly, and shortly afterwards found myself in the
comfortable halls of Croxteth.

The promise, then, with which I set out, and which
many gentlemen, who pretended to know, said that I could
not keep, was kept.   In little over three months I had
crossed the Atlantic twice, had travelled by rail and
steamboat more than 1000 miles beyond New York, had
travelled the desert for a month in my waggon, and had
hunted, killed, and brought back to England the bison,
or the largest game that haunts the Plains of the Far
West.

I will now take this opportunity of referring to that
most useful little volume published by my friend Capt.
Marcy, of the United States Army, and entitled " The
Prairie Traveller."   In that work it will be seen—and no
one essaying a visit to the plains should be without it—
that it is not advisable to attempt the prairies, or to sojourn
or hunt in the Indian grounds, unless associated with a
force of fifty men.   Of this of course he is a better judge
than I can be, if experience is deemed to be the teacher ;

but, as my finances did not permit me to hire many men, and time and circumstances did not enable me to associate myself with personal friends, why, I went to the plains with ten servants, nine of whom were hired on the edge of the desert, many of whom could scarcely be depended on, and all of whom, except the servant I brought with me, were badly or insufficiently armed ; at a time, too, when the Indians had assumed their war paint, and were attacking all the white men that came within the scope of their revenge. This on account of the death of the petty Indian chief, Pawnee, whom Mr Bayard, of the United States Army, had slain in the strict performance of his military duty.

As far as my short experience goes, the chief danger to be apprehended at the hands of the Indians is a surprise. Like the tiger, they will not attack a man unless they think they have him at advantage, and they are terribly afraid of the rifle in white men's hands. From all I could learn, the shooting of the red men is a farce as compared with that of the whites, and of course their fire-arms, as well as their ammunition, are for the most part damaged or in very inferior order. The bow and arrow, as used by savages, is in England much overrated; novelists, in regard to the rifle and the bow in red hands, as well as with reference to the personal appearance and habits of the savages, have beautifully gulled their readers; nor has Mr Murray in any way fallen short of inflammatory description, when in the book he published of his travels he so much descanted on the romantic pleasure of his association and life on the prairies with the Pawnees. As my excellent friend—now, alas! no more—the late Lord Kennedy, used to say, when a man told a most wonderful thing that he had done, that " he would bet him ten thousand to one he would

never do it again ;" so with Mr Murray's tale of his sojourn
with the Pawnees on the prairies, eating raw bison, liver,
&c. &c. From my knowledge of what that tribe at present
is, I would bet him similar immense odds that he " never
did it again," nor any other gentleman breathing, had
he sojourned with them. I think it must have been some
other tribe; for such are the degraded and thievish propensities of that abominably filthy and squalid race, that
any white man would not only have been robbed and
murdered, but ere his death he would most assuredly
have had to submit his head to the will, or at least
to the inspection, of the chiefs on sunny days, in
their camps, to have afforded them the pastime of refresh-
ment, entomological capture, and food, for the Pawnees
are the only tribe who deem the most revolting insects a
luxury. It is a favourite occupation of the chiefs of the
Pawnees in idle hours—and they are always idle—to order
the children of the whole tribe thus to contribute heads
full of filth to the enjoyment of their elders.

But to return to Captain Marcy's book, " The Prairie
Traveller."

In one point alone my gallant friend and myself considerably differ, and that is in the best method of swimming the horse. Capt. Marcy says, in page 79 of his
book, and with reference to crossing rivers in a swollen
state, " If the traveller be alone, his only way is to swim
his horse; but if he retains his seat in his saddle, his
weight presses the animal down into the water and
cramps his movements very sensibly. It is a much
better plan to attach a cord to the bridle-bit and drive him
into the stream, then, seizing his tail, allow him to tow you
across." Now, in this advice I cannot agree; whatever
a man has to do with a horse with a saddle on his back,

let the man never abdicate his throne, but stick to his
pig-skin as he would to his ship, with all his wits about
him. It is very dangerous to lay hold of the tail of a
horse, and to let him tow you in the water, and for this
reason,—in swimming, a horse uses his hinder legs
much as if he were in full trot, and having no ground to
catch the impetus of the heels, which all the time seek
and expect to be stayed by it, the hind legs go forth as it
were in a succession of kicks, and to the very extent of
their muscular power. If one of these kicks happened
to strike a man, he would be injured and probably
drowned. If a man must be off his saddle in the water
on account of the inferiority or weakness of his horse, let
him hold as lightly as he can to the mane, and thus
support himself and maintain command of the animal,
but on no account get behind him. I have swum by the
side of a horse, and I know very well what I am saying.
In the picture accompanying the 79th page, wherein
Captain Marcy and myself differ, the illustration of a horse
thus swimming is not at all what it should be, and it is
not in accordance with the writer's instructions as to the
"loose rein." The rein of the swimming horse in the
picture is too tight, and the rein, unless a rider sat on
his back to loosen it sufficiently, should never be left in
that position on a horse when in the water. The rein on
an unridden horse should be gathered up to the check in
such a way that it could neither get under the foreleg of
the horse, nor have any action on his bit. The least touch
of the mouth guides a horse when in the water ; the least
touch stops, and, if over-hard, sets him fighting and pawing, and even, when the water is very deep, brings him
backwards, and risks his life. The stirrups too, if a
man is not on his horse's back, should be crossed safely over

the saddle ; and in swimming horses with men on their backs, even then I recommend the stirrups being taken out of the water, and the riders should sit with their knees well forward, and their legs gathered up as far as the firmness of the seat will permit, to prevent any collision with the feet of the horse. In guiding a horse from the saddle in the water, the touch to the mouth on either side to direct him should be very light and entirely made by the rein on that side, and if the horse has a double bit in his mouth the direction should be entirely confined to the snaffle. In the swimming of different horses there is even more difference in their action than there is in that of dogs ; some will swim as smoothly on the surface as a dog, while others will go through a river by a succession of plunges to the bottom, if they can touch it, and up again, and these are difficult to sit and manage, for as they plunge under the water they come up half blinded and shaking their heads from the water in their ears, so that they are almost out of their senses. With the difference as to the swimming of horses — I believe it is the only one between Capt. Marcy and myself—I now take leave of this excellent little book, reiterating my opinion that no traveller should attempt the plains without first having studied the volume, or possessed himself of it, so that he could have it at command at any needful moment.

Of all routes for seeking the bison, the following is infinitely the best. On reaching New York, proceed thence to Niagara, and there put up at the Monteagle Hotel. A day or more can well be spent in viewing the splendid Falls, and then the route which I have described, taken thus the reverse way from which I travelled it, will, in passing, place at the traveller's option lakes and rivers for fishing or shooting, and the hotel, the society, and the yacht, upon the lake Michigan, of Mr Fellers. In him the traveller will find a sportsman and gentleman and meet with all that attention and kindness which render life in any place so very agreeable. In the route by which I returned, the traveller will find St Louis, and there let him deliver a sufficient letter of introduction to the Messrs Campbell. At St Louis let him inquire for my friend Captain Lousley (at the time that I was there in command of the steamer Skylark), and let him thus wend his way by that steamer far as he can go, to St Joe. At St Joe let him inquire for any of the gentlemen named in my narrative, and, having delivered a letter of introduction to them from the Messrs Campbell, perhaps they, if they like his manners and looks, and he pretends to no vulgar eccentricity, will make a sufficient party and join him in a hunt on the plains. If this cannot be done, let the traveller then take the cumbrous, but for a public conveyance not unpleasant, vehicle, called a four-horse stage-coach, and let him jolt his way in this to Fort Riley. Supposing him to have sufficient letters of introduction to the gentlemen and soldiers of the American Army there in command, he will meet with as high-hearted, generous hospitality as any place in England could afford, and they will, if their duty permits them to do so, form a party that will, in safety and comfort and good companionship, take him to the plains, and good-humouredly test his hunting powers. By this means all travellers will get rid of the nuisance of hiring a guide, who may be, as mine was, a rascal, incompetent to anything but falsehood, disobedience, and theft. They will save an immense expenditure in wages, they need buy no mules, but they will have to run their chance of picking up a horse or two, for the purpose of hunting, at St Joseph or Fort Riley ; and these they sell again when the chase is over.

Adieu, then, to the United States, to her large and scrambling cities, which are really but beginning to lick themselves into shape. To begin with the beginning, the greater portion of the streets of New York are miserably paved, and grass grows on many of its flagged footways, while in the back slums or smaller streets drainage is not thought of, and decayed fruit and vegetables, rinds of melons and potatoes, cabbage leaves, and other rotting horrors, lie steaming on the gutterless side of the pavement, and sending the fumes of cholera into the dark and crowded recesses of the confined cottage. A few pretty churches, one noble street, and a decent square or two, comprise the perfections of the architecture of New York. There are no cathedrals, no fine mansions, and little to recommend the rest of the town, save its facilities for the passing of people and property across the different ferries on the river. The railways of America are horrible, as at present managed ; and there is nothing to recommend them for but the length to which they are now running. The steam navigation on the river is good, and far superior to the conduct of the rail, and common and ignorant report has been more unjust to the captains in command of the river steamers ; they are an able, a steady, and an excellent class of men. The best class of American society is as good as any society ; the ladies are charming, and the gentlemen are in mind high-toned and urbane. The tradesmen are for the most part assuming, uncivil, and under a lamentable mistake in the supposition that rudeness to their betters proves that there are no betters, but that a general equality pervades the land. Of course to this there are exceptions, and when you do meet with a civil tradesman the contrast to the generality is so great that in spite of yourself you buy of him more than you require. The Boh-hoys, or blackguards, their name in any town is legion, and in place of a few peers and commoners verbally dictating to their tenants for their votes at elections,—in America you have an irresponsible multitude, who have not a real or vested interest in the State, coercing with knives, rifles, and revolvers the free exercise of the franchise. Entailing on the representation of the people a false position and a universal grievance, which will one day be the severance of the States, and end in anarchy and confusion.

The intensity of heat and cold arrives at like results, and no mind ever yet contemplated the histories of countries without arriving at the conclusion that a dominant and a grinding aristocracy, or a too free people, alike end in rearing up a tyrannical dictator to a throne, that is absolutely necessary for a time to curb licentiousness, and to weed the world of a wild growth, incompatible with the life of utility and the interests of civilisation. The plains or deserts of America are beautiful, and her wild creatures, except the Red Men, interesting and grand. The soil in the prairies is oftentimes so good that they may be said to afford remunerative space enough for agriculture to run wild on, before fields can be fenced in, or toil know where to rest from the occupation of the hour. The mineral fields the same; they hold the riches of *an* earth, and a thousand ages, and generations on generations, will die out, ere half the riches entombed beneath the soil are fused and diffused, for the good or otherwise of mankind. What then shall I say of the

general press in the United States, the established voice
of the land, which ought to lead to truth and virtue, to
set an example of right, and to hold up all wrongs to
condemnation; to forbear from unfairly attacking a
stranger, and never to be induced to the use of language
which would scarcely be uttered face to face. The pen
is in my hand now, but evil example shall never induce me
to forget the gentlemanly and courteous usages of English
society, and thanking that portion of the American press
that has done me justice and given me fair play, I for-
give some trifling nationalities among a few of my transatlantic brethren, which did their country no good, did
me no harm, and served only to support the views of
" Boz."

In thus arriving at the end of my narrative, it is curious and yet painful to me to see that, if " not in my own
country," still in America, while I was there I was " a
true prophet," for in my lectures given at St Joseph
and St Louis, and since in my writings published in the
English press, I warned the inhabitants of the United States
that the Union was in danger. Gallant and sincere friends
of mine, among them Col. Sumner, high in the standing
army, combatted my impressions in this matter, but
what have become of their arguments now? The storm,
that must ever gather on a political horizon swayed by
universal suffrage and darkened by selfish demagogues,
who guide the masses to no common, no other interest
than that of the ephemerally raised leader, is now on
the eve of bursting, and, as far as man can see, the slave
question is shaking America to her very centre, and the
" strength of Union" about to be lost. A lesson, a terrible lesson, I fear, to America, and a useful one to England, is about to be read. Heaven grant that it may be

averted; still, if come it must, let not the Old Country any more be told by demagogues at home that she ought to copy from the political institutions of the Younger Land, for all the danger now brewing for the severance of unity in the United States, arises from an interference with property by the masses and an overwhelming democracy; from the want of an educated Congress fairly representing state, station, and the vested wealth of the country; and from an insane love of liberty, which when carried too far ever becomes licentiousness, and ends in anarchy and confusion. That America may weather the storm and ward off all difficulties, must ever be the sincere wish of

THE AUTHOR.

THE END.

NEW AND INTERESTING WORKS

PUBLISHED BY

MESSRS. HURST AND BLACKETT,

SUCCESSORS TO MR. COLBURN.

MEMOIRS OF THE COURTS AND CABINETS OF

WILLIAM IV. AND VICTORIA. FROM ORIGINAL FAMILY DOCU-
MENTS. By the DUKE OF BUCKINGHAM AND CHANDOS, K.G.
Completing the BUCKINGHAM PAPERS. 2 vols. 8vo. with Portraits. 30s.

Among the principal interesting subjects of these volumes will be found :—
The Re-establishment of the Royal Household—The Sailor King and his Court
—The Duke of Wellington In, and Out of, Office—The Reform Cabinet and the
Conservative Opposition—Career of Sir Robert Peel—Civil List Expenditure—
Vicissitudes of Louis Philippe—Attacks on the Duke af Wellington—Corona-
tions of William IV. and Queen Victoria—Rise and Fall of O'Connell—Lord Mel-
bourne and his Ministry—Proceedings of the Kings of Hanover and Belgium—Pri-
vate Negotiations at Apsley House—Secret History of Court Arrangements. &c.

"These volumes bring to a conclusion the interesting series of memoirs which have been
published under the auspices of the Duke of Buckingham during the last few years.
Founded on the traditions of a family whose members have long possessed the *entrée* into
the charmed circle of courtiers and politicians, and enriched by the private and confidential
letters of the great men of the time, these works possess a peculiar interest which is not
always the attribute of state memoirs. They lift the veil of mystery with which the agents
of court influence and cabinet intrigues shroud their actions from the eyes of the public
and show us the motives which actuated our statesmen, and the degree in which the private
expressions of their views coincided with the public declaration of th ir sentiments. The
number of original documents in the present volumes invests the work with a fresh and
authentic interest. As forming the conclusion of a valuable and important series, these
memoirs should find a place on the shelves of every library."—*Sun.*

MEMOIRS OF THE COURT OF GEORGE IV. FROM

ORIGINAL FAMILY DOCUMENTS. By the DUKE OF BUCKINGHAM
AND CHANDOS, K.G. 2 vols. 8vo. with Portraits. 30s. bound.

"The country is very much indebted to the Duke of Buckingham for the publication of
these volumes—to our thinking the most valuable of the contributions to recent history
which he has yet compiled from his family papers. Besides the King, the Duke of
Buckingham's canvass is full of the leading men of the day—Castlereagh, Liverpool, Canning, Wellington, Peel, and their compeers. We are sure that no reader, whether he seeks
for gossip, or for more sterling information, will be disappointed by the book. There are
several most characteristic letters of the Duke of Wellington."—*John Bull.*

"The original documents and private letters published in these volumes—penned by
public men, who were themselves active participators in the events and scenes described
—throw a great deal of very curious and very valuable light upon this period of our history.
Written in the absence of all restraint, they necessarily possess a high interest even
for the lightest and most careless reader ; whilst, in an historical sense, as an authentic
source from which future historians will be enabled to form their estimate of the
characters of the leading men who flourished in the reign of the last George, they must
be regarded as possessing an almost inestimable value. Taking this publication altogether,
we must give the Duke of Buckingham great credit for the manner in which he has
executed it, and at the same time return him our hearty thanks for the interesting and
valuable information which he has unfolded to us from his family archives."—*Observer.*

MEMOIRS OF THE COURT OF THE REGENCY.

FROM ORIGINAL FAMILY DOCUMENTS. By the DUKE OF BUCKING-
HAM AND CHANDOS, K.G. 2 vols. 8vo., with Portraits, 30s. bound.

"Here are two more goodly volumes on the English Court; volumes full of new
sayings, pictures, anecdotes, and scenes. The Duke of Buckingham travels over nine years
of English history. But what years those were, from 1811 to 1820! What events at home
and abroad they bore to the great bourne!—from the accession of the Regent to power to
the death of George III.—including the fall of Perceval; the invasion of Russia, and the
war in Spain; the battles of Salamanca and Borodino; the fire of Moscow; the retreat of
Napoleon; the conquest of Spain; the surrender of Napoleon; the return from Elba; the
Congress of Vienna; the Hundred Days; the crowning carnage of Waterloo; the exile to
St. Helena; the return of the Bourbons; the settlement of Europe; the public scandals a
the English Court; the popular discontent, and the massacre of Peterloo! On many parts
of this story the documents published by the Duke of Buckingham cast new jets of light,
clearing up much secret history. Old stories are confirmed—new traits of character are
brought out. In short, many new and pleasant additions are made to our knowledge of
those times."—*Athenæum.*

"Invaluable, as showing the true light in which many of the stirring events of the
Regency are to be viewed. The lovers of Court gossip will also find not a little for their
edification and amusement."—*Literary Gazette.*

"These volumes cover a complete epoch, the period of the Regency—a period of large
and stirring English history. To the Duke of Buckingham, who thus, out of his family
archives, places within our reach authentic and exceedingly minute pictures of the governors
of England, we owe grateful acknowledgements. His papers abound in fresh lights on old
topics, and in new illustrations and anecdotes. The intrinsic value of the letters is enhanced
by the judicious setting of the explanatory comment that accompanies them, which is put
together with much care and honesty."—*Examiner.*

HISTORY OF THE REIGN OF HENRY IV., KING OF

FRANCE AND NAVARRE. From numerous Original Sources. By MISS
FREER. Author of "The Lives of Marguerite d'Angoulême, Elizabeth
de Valois, Henry III." &c. 2 vols. with Portraits, 21s.

"Various circumstances combine to make us regard the Life of Henry IV. as one of the
most attractive in the wide range of biography. The chequered nature of his career from
childhood to manhood, the perils that environed him in a Court hostile to his religion and
race, his unfortunate marriage, his personal bravery, his skill as a commander—these and
many other characteristics that will suggest themselves to our readers, cause us to hail
Miss Freer's new work as a welcome addition to our stock of books. It is a well-known
feature in Miss Freer's works, that not content with the ordinary sources of information to
which popular writers have recourse, she investigates for herself the MS. documents of the
period under review, and is thus enabled to supply us with new facts, and to bring us face
to face with the persons whose actions are recorded. This, which constitutes one of the
great charms of M. Michelet, as a historian, is likewise a marked characteristic of Miss
Freer, and confers a great additional value upon her historical portraits."—*Critic.*

"To become the chronicler of such a reign as that of Henry IV. is no mean task, and
Miss Freer has accomplished it with singular good taste, good sense, and vigour. The
story never flags. Our authoress is always faithful, accurate, and intelligent. Her style
is good, and her subject abounds with interest for every student of history."—*Herald.*

"We know no works of this kind, with the exception, perhaps, of Macaulay's history,
which are more pleasant reading than the histories of Miss Freer. The charm of the style
and manner, and the accuracy of the details, combine to render her works a valuable
addition to our literary treasures."—*John Bull.*

"In telling the reign of Henry IV., Miss Freer has one of the most interesting portions
of French history for her story. She has told it from first to last with taste, using a clear,
vigorous style."—*Examiner.*

"The public will thank Miss Freer most heartily for these delightful volumes. In her
particular line she is the best historian of her day."—*Chronicle.*

HENRY III. KING OF FRANCE AND POLAND;

HIS COURT AND TIMES. From numerous unpublished sources, including MS. Documents in the Bibliothèque Impériale, and the Archives of France and Italy. By MISS FREER, Author of " Marguerite d'Angoulême," " Elizabeth de Valois, and the Court of Philip II," &c. 3 vols. post 8vo. with fine portraits, 31s. 6d. bound.

" Miss Freer having won for herself the reputation of a most painstaking and trustworthy historian not less than an accomplished writer, by her previous memoirs of sovereigns of the houses of Valois and Navarre, will not fail to meet with a most cordial and hearty welcome for her present admirable history of Henry III., the last of the French kings of the house of Valois. We refer our readers to the volumes themselves for the interesting details of the life and reign of Henry III., his residence in Poland, his marriage with Louise de Lorraine, his cruelties, his hypocrisies, his penances, his assassination by the hands of the monk Jaques Clément, &c. Upon these points, as well as with reference to other persons who occupied a prominent position during this period, abundant information is afforded by Miss Freer; and the public will feel with us that a deep debt of gratitude is due to that lady for the faithful and admirable manner in which she has pourtrayed the Court and Times of Henry the Third."—*Chronicle.*

" The previous historical labours of Miss Freer were so successful as to afford a rich promise in the present undertaking, the performance of which, it is not too much to say, exceeds expectation, and testifies to her being not only the most accomplished, but the most accurate of modern female historians. The Life of Henry III. of France is a contribution to literature which will have a reputation as imperishable as its present fame must be large and increasing. Indeed, the book is of such a truly fascinating character, that once begun it is impossible to leave it."—*Messenger.*

" Among the class of chronicle histories, Miss Freer's Henry the Third of France is entitled to a high rank. As regards style and treatment Miss Freer has made a great advance upon her 'Elizabeth de Valois,' as that book was an advance upon her 'Marguerite D'Angoulême.' "—*Spectator.*

" We heartily recommend this work to the reading public. Miss Freer has much, perhaps all, of the quick perception and picturesque style by which Miss Strickland has earned her well-deserved popularity."—*Critic.*

ELIZABETH DE VALOIS, QUEEN OF SPAIN, AND

THE COURT OF PHILIP II. From numerous unpublished sources in the Archives of France, Italy, and Spain. By MISS FREER. 2 vols post 8vo. with fine Portraits by HEATH. 21s.

" It is not attributing too much to Miss Freer to say that herself and Mr. Prescott are probably the best samples of our modern biographers. The present volumes will be a boon to posterity for which it will be grateful. Equally suitable for instruction and amusement, they portray one of the most interesting characters and periods of history."—*John Bull.*

" Such a book as the memoir of Elizabeth de Valois is a literary treasure which will be the more appreciated as its merits obtain that reputation to which they most justly are entitled. Miss Freer has done her utmost to make the facts of Elizabeth's, Don Carlos', and Philip II.'s careers fully known, as they actually transpired."—*Bell's Messenger*

THE LIFE OF MARGUERITE D'ANGOULEME.

QUEEN of NAVARRE, SISTER of FRANCIS I. By MISS FREER. Second Edition, 2 vols. with fine Portraits, 21s.

" This is a very useful and amusing book. It is a good work, very well done. The authoress is quite equal in power and grace to Miss Strickland. She must have spent great time and labour in collecting the information, which she imparts in an easy and agreeable manner. It is difficult to lay down her book after having once begun it. This is owing partly to the interesting nature of the subject, partly to the skilful manner in which it has been treated. No other life of Marguerite has yet been published, even in France. Indeed, till Louis Philippe ordered the collection and publication of manuscripts relating to the history of France, no such work could be published. It is difficult to conceive how, under any circumstances, it could have been better done."—*Standard.*

LODGE'S PEERAGE AND BARONETAGE FOR 1861.

UNDER THE ESPECIAL PATRONAGE OF HER MAJESTY AND H.R.H. THE
PRINCE CONSORT. Corrected throughout by the Nobility. THIRTIETH
EDITION, in 1 vol. royal 8vo., with the Arms beautifully engraved, hand-
somely bound, with gilt edges, price 31s. 6d.

LODGE'S PEERAGE AND BARONETAGE is acknowledged to be the most
complete, as well as the most elegant, work of the kind. As an established and
authentic authority on all questions respecting the family histories, honours,
and connections of the titled aristocracy, no work has ever stood so high. It is
published under the especial patronage of Her Majesty, and His Royal Highness
the Prince Consort, and is annually corrected throughout, from the personal
communications of the Nobility. It is the only work of its class, in which,
the type being kept constantly standing, every correction is made in its proper
place to the date of publication, an advantage which gives it supremacy over all
its competitors. Independently of its full and authentic information respecting
the existing Peers and Baronets of the realm, the most sedulous attention is
given in its pages to the collateral branches of the various noble families, and
the names of many thousand individuals are introduced, which do not appear in
other records of the titled classes. For its authority, correctness, and facility of
arrangement, and the beauty of its typography and binding, the work is justly entitled to the high place it occupies on the tables of Her Majesty and the Nobility.

"Lodge's Peerage must supersede all other works of the kind, for two reasons; first, it
is on a better plan; and, secondly, it is better executed. We can safely pronounce it to be
the readiest, the most useful, and exactest of modern works on the subject."—*Spectator.*

"A work which corrects all errors of former works. It is the production of a herald,
we had almost said, by birth, but certainly by profession and studies, Mr. Lodge, the Norroy
King of Arms. It is a most useful publication."—*Times.*

"As perfect a Peerage of the British Empire as we are ever likely to see published.
Great pains have been taken to make it as complete and accurate as possible. The work
is patronised by Her Majesty and the Prince Consort; and it is worthy of a place in every
gentleman's library, as well as in every public institution."—*Herald.*

"As a work of contemporaneous history, this volume is of great value—the materials
having been derived from the most authentic sources and in the majority of cases emanating
from the noble families themselves. It contains all the needful information respecting the
nobility of the Empire."—*Post.*

"This work should form a portion of every gentleman's library. At all times, the information which it contains, derived from official sources exclusively at the command of the
author, is of importance to most classes of the community; to the antiquary it must be
invaluable, for implicit reliance may be placed on its contents."—*Globe.*

"This work derives great value from the high authority of Mr. Lodge. The plan
is excellent."—*Literary Gazette.*

"When any book has run through so many editions, its reputation is so indelibly
stamped, that it requires neither criticism nor praise. It is but just, however, to say, that
'Lodge's Peerage and Baronetage' is the most elegant and accurate, and the best of its
class. The chief point of excellence attaching to this Peerage consists neither in its
elegance of type nor its completeness of illustration, but in its authenticity, which is insured
by the letter-press being always kept standing, and by immediate alteration being made
whenever any change takes place, either by death or otherwise, amongst the nobility of the
United Kingdom. The work has obtained the special patronage of Her Most Gracious
Majesty, and of His Royal Highness the Prince Consort, which patronage has never been
better or more worthily bestowed."—*Messenger.*

"'Lodge's Peerage and Baronetage' has become, as it were, an 'institution' of this
country; in other words, it is indispensable, and cannot be done without, by any person
having business in the great world. The authenticity of this valuable work, as regards the
several topics to which it refers, has never been exceeded, and, consequently, it must be
received as one of the most important contributions to social and domestic history extant.
As a book of reference—indispensible in most cases, useful in all—it should be in the
hands of every one having connections in, or transactions with, the aristocracy."—*Observer.*

LODGE'S GENEALOGY OF THE PEERAGE AND

BARONETAGE OF THE BRITISH EMPIRE. A New and Revised Edition. Uniform with "The Peerage" Volume, with the arms beautifully engraved, handsomely bound with gilt edges, price 31s. 6d.

The desire very generally manifested for a republication of this volume has dictated the present entire revision of its contents. The Armorial Bearings prefixed to the History of each Noble Family, render the work complete in itself and uniform with the Volume of The Peerage, which it is intended to accompany and illustrate. The object of the whole Work, in its two distinct yet combined characters, has been useful and correct information; and the careful attention devoted to this object throughout will, it is hoped, render the Work worthy of the August Patronage with which it is honoured and of the liberal assistance accorded by its Noble Correspondents, and will secure from them and from the Public, the same cordial reception it has hitherto experienced. The great advantage of "The Genealogy" being thus given in a separate volume, Mr. Lodge has himself explained in the Preface to "The Peerage."

MEMORIALS OF ADMIRAL LORD GAMBIER, G.C.B.

with Original Letters from Lords Chatham, Nelson, Castlereagh, Mulgrave, Holland, Mr. Canning, &c, Edited, from Family Papers, by Lady CHATTERTON, Second Edition, 2 vols. 8vo, 28s.

"Lady Chatterton is not only a zealous but a skilful biographer. These volumes are among the most readable as well as most important books of the season."—*Observer.*

"These volumes are an important addition to our naval literature; but they are also valuable for the light they throw on the domestic history of the time. The correspondence is particularly rich in anecdotes, glimpses of society and manners, and traits of character."—*U. S. Magazine.*

"An important and valuable addition to the history of Lord Gambier's times."—*Messenger.*

A BOOK ABOUT DOCTORS. BY J. C. JEAFFRESON.

Esq., Author of "Novels and Novelists," &c. 2 vols. with plates. 21s.

"This is a rare book; a compliment to the medical profession and an acquisition to its members; a book to be read and re-read; fit for the study and the consulting-room, as well as the drawing-room table and the circulating library. Mr. Jeaffreson takes a comprehensive view of the social history of the profession, and illustrates its course by a series of biographic and domestic sketches, from the feudal era down to the present day. The chapters on the Doctor as a bon-vivant, the generosity and parsimony, the quarrels and loves of physicians, are rich with anecdotes of medical celebrities. But Mr. Jeaffreson does not merely amuse. The pages he devotes to the exposure and history of charlatanry are of scarcely less value to the student of medicine than the student of manners. We thank Mr. Jeaffreson most heartily for the mirth and solid information of his volumes. They appeal to a wide circle. All the members of our profession will be sure to read them."—*Lancet.*

"A pleasant book for the fireside season on which we are now entering, and for the seaside season that is to come. Out of hundreds of volumes, Mr. Jeaffreson has collected thousands of good things, adding much that appears in print for the first time, and which of course gives increased value to this very readable book."—*Athenæum.*

DOMESTIC MEMOIRS OF THE ROYAL FAMILY,

and the COURT OF ENGLAND, chiefly at SHENE and RICHMOND.

By FOLKESTONE WILLIAMS, F.G.S.,   3 vols. Portraits.

"In the prosecution of his labours, the author has consulted antiquaries and archæologists, and examined contemporary authorities. The result is, a work, pleasant and instructive, abundant in anecdote, and agreeably gossipping. It, moreover, evinces considerable research, and a generally sound historical judgment."—*Spectator.*

"This work belongs to the best class of popular antiquarian books, because it is popular by reason of the entertaining character and the variety of its store of trust-worthy information."—*Examiner.*

ESSAYS FROM THE QUARTERLY.   BY JAMES

HANNAY. 1 vol. 8vo. (Just ready).

THE LIFE AND TIMES OF GEORGE VILLIERS,

DUKE OF BUCKINGHAM. By MRS. THOMSON. 3 vols.

"These volumes will increase the well-earned reputation of their clever and popular author. The story of the royal favourite's career is told by Mrs. Thomson very honestly, and is enriched abundantly with curious and entertaining details from the familiar letters of the time and the memorials of the State Paper Office, of which a full publication is now made for the first time. Labour and pains have, indeed, been well spent upon volumes that produce their evidence so fairly and are written so agreeably as these."—*Examiner.*

"Mrs. Thomson is entitled to great praise. She has written the most complete biography of Buckingham that has appeared in the language. Those who commence the work by being amused will end in being instructed."—*Literary Gazette.*

BRITISH ARTISTS, from HOGARTH to TURNER;

Being a SERIES OF BIOGRAPHICAL SKETCHES. By WALTER THRONBURY. 2 vols. 21s.

"Mr. Thornbury writes with knowledge and enthusiasm. The interest of his sketches is unquestionable."—*Examiner.*

"The interest of Mr. Thornbury's pictures is undeniable—a result partly due to the talent of the painter, partly to his subjects; for next to the lives of actors those of artists are among the most interesting to read. Especially so are those of our English artists of the last century—lives abounding in contrasted and often dark hues, interwoven with the history of men still remarkable in letters and politics. Capital subjects for a biographer with a turn for dramatic and picturesque realisation are such men as the bright, mercurial Gainsborough; the moody, neglected Wilson; Reynolds, the bland and self-possessed; Barry, the fierce and squalid; shrewd, miserly Nollekins; the foppish, visionary Conway; the spendthrift Sherwin; the stormy Fuseli; Morland, the reprobate; Lawrence, the courtly. The chapters devoted to these heroes of the English schools are not so much condensed biographies as dramatic glimpses of the men and their environments. Certain striking scenes and circumstances in their lives are vividly and picturesquely painted—made to re-live before our eyes with all the vraisemblance and illusive effect of the novelist."—*Critic.*

"Mr. Thornbury's delightful artistic sketches will be gladly welcomed. Graphic in design and brilliant in style, the work can scarcely fail to find favours with the lovers of British literature as well as of British art."—*Sun.*

MEMOIRS OF ROYAL LADIES.   BY EMILY S.

HOLT. 2 volumes post 8vo. with Illustrations. (Just ready.)

TRAVELS IN THE REGIONS OF THE AMOOR,

AND THE RUSSIAN ACQUISITIONS ON THE CONFINES OF INDIA AND
CHINA; WITH ADVENTURES AMONG THE MOUNTAIN KIRGHIS, AND THE
MANJOURS, MANYARGS, TOUNGOUZ, TOUZEMTZ, GOLDI, AND GELYAKS.
By T. W. ATKINSON, F G.S., F.R.G.S., Author of " Oriental and Western
Siberia." Dedicated by permission, to HER MAJESTY. SECOND EDITION.
Royal 8vo., with Map and 83 Illustrations. £2 2s., elegantly bound

" Our readers have not now to learn for the first time the quality of Mr. Atkinson as an
explorer and a writer. The comments we made on, and the extracts we selected from, his
' Oriental and Western Siberia' will have sufficed to show that in the former character he
takes rank with the most daring of the class, and that in the latter he is scarcely to be
surpassed for the lucidity, picturesqueness, and power, with which he pourtrays the scenes
through which he has travelled, and the perils or the pleasures which encountered him on
the way. The present volume is not inferior to its predecessor. It deals with civilization,
semi-civilization, and barbarous life. It takes us through localities, some of which are
little, others not at all, known to even the best read men in the literature of travel. The
entire volume is admirable for its spirit, unexaggerated tone, and the mass of fresh materials
by which this really new world is made accessible to us. The followers, too, of all the ' ologies'
will meet with something in these graphic pages of peculiar interest to them. It is a noble
work."—Athenæum.

" We must refer to Mr. Atkinson as one of the most intelligent and successful of the
civilized travellers of our own day. By far the most important contribution to the history
of these regions is to be found in Mr. Atkinson's recent publication on the Amoor—a work
which derives equal interest from his well-stored portfolio and his pen."—Edinburgh
Review.

" This is in every respect an aureus liber. Its magnificent apparel not inaptly sym-
bolises its magnificent contents. Mr. Atkinson has here given us a narrative which could
be told by no other living Englishman. The intrinsic interest of that narrative is enhanced
by Mr. Atkinson's gift of vigorous and graceful description. Thanks to the power of his
pen, and the still more remarkable power of his pencil we follow his travels with eager
interest and anxiety. He himself is the chief object of interest, from his thirst for adventure
and daring exploits, and the countless shapes of terror and death that he encounters.
The work is a magnificent contribution to the literature of travel. More useful and
pleasant reading can no where be found."—Literary Gazette.

" Mr. Atkinson has here presented the reading world with another valuable book of
travels. It is as interesting, as entertaining, and as well written as his previous work. It
is a volume which will not only afford intellectual entertainment of the highest order, but
fitted to instruct both the philosopher and the statesman. The vast territorial acquisitions
lately made by Russia in the Northern parts of Central Asia along the whole frontier of
China, is described by an eye wi ness well qualified to estimate their real value and political
advantages. Our readers, we feel sure, will peruse this interesting book of travels for
themselves. It contains something for every taste."—Daily News

" The success of Mr. Atkinson's ' Oriental and Western Siberia' has happily induced
him to write and publish another volume, and written with the same unflagging interest.
A more pleasing as well as more novel book of travels it would be difficult to find. The
illustrations are admirably executed, and they add ten fold to the value of a volume already
possessing intrinsic merits of the highest kind. Independently of the deep interest it excites
as a traveller's tale, the work has other claims. It presents peculiar geographical and ethnolo-
gical information, and points out a boundless field of commerce to English enterprise. It
marks with a decided pen the gradual advance of Russia towards British India, and the
sweeping rush of her conquering energy from Siberia to the Pacific. Thus Mr. Atkinson's
book has not only a literary, but a political and commercial importance. There is food for
all readers and interest for all."—Globe.

" This is noble and fascinating book, belonging in right both of subject and treatment
to the choicest class of travel literature. The vast panorama unfolded is one of the most
marvellous in the world, and has hitherto been among the least known to th e nations of
the west. It is now set before them with exquisite clearness and force of expression by one
who has the highest claims to confidence as an observer and delineator."—Spectator.

" A really magnificent volume, which for many years to come must be a standard
authority upon the country of which it treats. It is very interesting and abounds in
incident and anecdote both personal and local."—Chronicle.

ORIENTAL AND WESTERN SIBERIA; A NAR-

RATIVE OF SEVEN YEARS' EXPLORATIONS AND ADVENTURES IN SIBERIA, MONGOLIA, THE KIRGHIS STEPPES, CHINESE TARTARY, AND CENTRAL ASIA. By THOMAS WITLAM ATKINSON. In one large volume, royal 8vo., Price £2. 2s., elegantly bound. Embellished with upwards of 50 Illustrations, including numerous beautifully coloured plates, from drawings by the Author, and a map.

"By virtue alike of its text and its pictures, we place this book of travel in the first rank among those illustrated gift books now so much sought by the public. Mr. Atkinson's book is most readable. The geographer finds in it notice of ground heretofore left undescribed, the ethnologist, geologist, and botanist, find notes and pictures, too, of which they know the value, the sportman's taste is gratified by chronicles of sport, the lover of adventure will find a number of perils and escapes to hang over, and the lover of a frank good-humoured way of speech will find the book a pleasant one in every page. Seven years of wandering, thirty-nine thousand five hundred miles of moving to and fro in a wild and almost unknown country, should yield a book worth reading, and they do."—*Examiner.*

"A book of travels which in value and sterling interest must take rank as a landmark in geographical literature. Its coloured illustrations and wood engravings are of a high order, and add a great charm to the narrative. Mr. Atkinson has travelled where it is believed no European has been before. He has seen nature in the wildest, sublimest, and also the most beautiful aspects the old world can present. These he has depicted by pen and pencil. He has done both well. Many a fireside will rejoice in the determination which converted the artist into an author. Mr. Atkinson is a thorough Englishman, brave and accomplished, a lover of adventure and sport of every kind. He knows enough of mineralogy, geology, and botany to impart a scientific interest to his descriptions and drawings; possessing a keen sense of humour, he tells many a racy story. The sportsman and the lover of adventure, whether by flood or field, will find ample stores in the stirring tales of his interesting travels."—*Daily News.*

"An animated and intelligent narrative, appreciably enriching the literature of English travel. Mr. Atkinson's sketches were made by express permission of the late Emperor of Russia. Perhaps no English artist was ever before admitted into this enchanted land of history, or provided with the talisman and amulet of a general passport; and well has Mr. Atkinson availed himself of the privilege. Our extracts will have served to illustrate the originality and variety of Mr. Atkinson's observations and adventures during his protracted wanderings of nearly forty thousand miles. Mr. Atkinson's pencil was never idle, and he has certainly brought home with him the forms, and colours, and other characteristics of a most extraordinary diversity of groups and scenes. As a sportsman Mr. Atkinson enjoyed a plenitude of excitement. His narrative is well stored with incidents of adventure. His ascent of the Bielouka is a chapter of the most vivid romance of travel, yet it is less attractive than his relations of wanderings across the Desert of Gobi and up the Tangnou Chain."—*Athenæum.*

"We predict that Mr. Atkinson's 'Siberia' will very often assume the shape of a Christmas Present or New Year's Gift, as it possesses, in an eminent degree, four very precious and suitable qualities for that purpose,—namely, usefulness, elegance, instruction and novelty. It is a work of great value, not merely on account of its splendid illustrations, but for the amount it contains of authentic and highly interesting intelligence concerning regions which, in all probability, has never, previous to Mr. Atkinson's explorations, been visited by an European. Mr. Atkinson's adventures are told in a manly style. The valuable and interesting information the book contains, gathered at a vast expense, is lucidly arranged, and altogether the work is one that the author-artist may well be proud of, and with which those who study it cannot fail to be delighted."—*John Bull.*

"To the geographer, the geologist, the ethnographer, the sportsman, and to those who read only for amusement, this will be an acceptable volume. Mr. Atkinson is not only an adventurous traveller, but a correct and amusing writer."—*Literary Gazette.*

SIX YEARS OF A TRAVELLER'S LIFE IN

WESTERN AFRICA. By FRANCISCO VALDEZ, Arbitrator at Loanda, and the Cape of Good Hope. 2 vols. with Illustrations.

" A book of value and importance. Its intrinsic merits are so many and so positive that we shall be greatly surprised if the work does not equal that by Dr. Livingstone in popularity, and be of similar value to the cause of Africa eventually."—*Messenger*.

"The immense amount of sterling information contained in this elaborate and instructive work cannot fail of ensuring a most cordial recognition of its merits at the hands of al such as have at heart the spread of Christianity and commerce in Africa."—*Lit. Gaz*

TWO YEARS IN SWITZERLAND AND ITALY.

By FREDRIKA BREMER. Translated by MARY HOWITT. 2 vols.

" A new work from the pen of Miss Bremer is ever hailed, not only with a hearty welcome, but with general acclamation. Such a reception will be given to this last specimen of her literary labours, which is certainly one of the best works she has ever yet produced. Where could such subjects as Switzerland and Italy find a more generous exponent? Who could appreciate the grandeur of the scenery of the land of freedom better than Fredrika Bremer? Who could see and understand all the phases of Italian society in its approaching struggle for liberty, better than this warm-hearted and generous woman? We have revelled in the volumes and can scarcely find words adequately to express our admiration of the manner in which Fredrika Bremer has told all she saw and felt during the two years she passed in the loveliest parts of Europe. The book is the best that ever was written on such themes."—*Messenger*.

TRAVELS IN EASTERN AFRICA, WITH THE

NARRATIVE OF A RESIDENCE IN MOZAMBIQUE: 1856 to 1859. By LYONS McLEOD, Esq. F.R.G.S., &c. Late British Consul in Mozambique. 2 vols. With Map and Illustrations. 21s.

"Mr. M'Leod's volumes contains chapters for all readers—racy narrative, abundance of incident, compendious history, important matter-of-fact statistics, and many a page which will be perused with pleasure by the naturalist."—*Athenæum*.

" Mr. M'Leod's work furnishes information concerning the commercial capabilities, not only of the Portugese settlements, but also of the Cape and Natal, together with particulars concerning Mauritius, Madagascar, and the Seychelles. It likewise gives a peculiar insight into the combinations and influences which operate upon the Portuguese authorities in relation to the slave trade."—*Times*.

LAKE NGAMI; OR EXPLORATIONS AND DIS-

COVERIES DURING FOUR YEARS' WANDERINGS IN THE WILDS OF SOUTH-WESTERN AFRICA. By CHARLES JOHN ANDERSSON. 1 vol. royal 8vo., with Map and upwards of 50 Illustrations, representing Sporting Adventures, Subjects of Natural History, &c. Second Edition.

"This narrative of African explorations and discoveries is one of the most important geographical works that have lately appeared. It contains the account of two journeys made between the years 1850 and 1854, in the first of which the countries of the Damaras and the Ovambo, previously scarcely known in Europe, were explored; and in the second the newly-discovered Lake Ngami was reached by a route that had been deemed impracticable, but which proves to be the shortest and the best. The work contains much scientific and accurate information as to the geology, the scenery, products, and resources of the regions explored, with notices of the religion, manners, and customs of the native tribes. The continual sporting adventures, and other remarkable occurrences, intermingled with the narrative of travel, make the book as interesting to read as a romance, as ,indeed, a good book of travels ought always to be. The illustrations by Wolf are admirably designed, and most of them represent scenes as striking as any witnessed by Jules Gérard or Gordon Cumming."—*Literary Gazette*.

A CRUISE IN THE PACIFIC: FROM THE LOG

OF A NAVAL OFFICER. Edited by CAPTAIN FENTON AYLMER. 2 v.

" A highly interesting work, written in the spirit of a genuine sailor."—*Lit. Gazette*.

THE MEDICAL MISSIONARY IN CHINA: A NAR-

RATIVE OF TWENTY YEARS' EXPERIENCE. By WILLIAM LOCK-
HART, F.R.C.S. F.R.G.S, of the London Missionary Society. Second
Edition, 1 vol. 8vo.

SEASONS WITH THE SEA HO SES; or, SPORTING

ADVENTURES IN THE NORTHERN SEAS. By JAMES LAMONT,
Esq. F.G.S. 8vo. with numerous Illustrations.

NARRATIVE OF A RESIDENCE AT THE COURT

OF MEER ALI MOORAD; WITH WILD SPORTS IN THE VALLEY OF
THE INDUS. BY CAPT. LANGLEY, late Madras Cavalry. 2 vols. 8vo.
with Illustrations. 30s.

"A valuable work, containing much useful information."—*Literary Gazette*.
"Captain Langley's interesting volumes will doubtless attract all the attention they
deserve on account of their political and commercial importance; and as they are full
of incident connected with the sports of British India, they will be as agreeable to the
sportsman and general reader as to the politician."—*Messenger*.

SIXTEEN YEARS OF AN ARTIST'S LIFE IN

MOROCCO, SPAIN, AND THE CANARY ISLANDS. By MRS.
ELIZABETH MURRAY. 2 vols. 8vo. with Coloured Illustrations.

"Mrs. Murray, wife, we believe, of the English Consul at Teneriffe, is one of the first of
female English Water Colour Artists. She draws well, and her colour is bright, pure, trans-
parent, and sparkling. Her book is like her painting, luminous, rich and fresh. We welcome
it (as the public will also do) with sincere pleasure. It is a hearty book, written by a clever,
quick-sighted, and thoughtful woman, who, slipping a steel pen on the end of her brush,
thus doubly armed, uses one end as well as the other, being with both a bright colourer,
and accurate describer of colours, outlines, sensations, landscapes and things. In a word,
Mrs. Murray is a clever artist, who writes forcibly and agreeably."—*Athenæum*.

A SUMMER RAMBLE in the HIMALAYAS; with

SPORTING ADVENTURES IN THE VALE OF CASHMERE. Edited
by MOUNTAINEER. 8vo. with Illustrations. 15s.

"This volume is altogether a pleasant one. It is written with zest and edited with care.
The incidents and adventures of the journey are most fascinating to a sportsman and very
interesting to a traveller."—*Athenæum*

SIX MONTHS IN REUNION: A CLERGYMAN'S

Holiday, and How he Passed it, By the Rev. P. BEATON, M.A. 2 v. 21s

"Mr. Beaton has done good service in the publication of these interesting volumes.
He is an intelligent observer, enjoys himself heartily, and compels his readers to enjoy
themselves also. Sagacity, practical good sense, a healthy animal nature, a well cultivated mind, are Mr. Beaton's qualifications as a traveller and a writer of travels. He
possesses the advantage, too, of having selected ground that is comparatively untrodden.
His work is written with taste and skill, and abounds with anecdote and information."
—*Literary Gazette*.

THE ENGLISHWOMAN IN ITALY: IMPRESSIONS

of Life in the Roman States and Sardinia, during a Ten Years' Residence.
By Mrs. G. GRETTON.

THE BOOK OF ORDERS OF KNIGHTHOOD, AND

DECORATIONS OF HONOUR OF ALL NATIONS; COMPRISING AN HISTORICAL ACCOUNT OF EACH ORDER, MILITARY, NAVAL AND CIVIL; with Lists of the Knights and Companions of each British Order. EMBELLISHED WITH FIVE HUNDRED FAC-SIMILE COLOURED ILLUSTRATIONS OF THE INSIGNIA OF THE VARIOUS ORDERS. Edited by SIR BERNARD BURKE, Ulster King of Arms. 1 vol. royal 8vo., handsomely bound, with gilt edges, price £2. 2s.

" This valuable and attractive work may claim the merit of being the best of its kind. It is so comprehensive in its character, and so elegant in style, that it far outstrips all competitors. A full historical account of the orders of every country is given, with lists of the Knights and Companions of each British Order. Among the most attractive features of the work are the illustrations. They are numerous and beautiful, highly coloured, and giving an exact representation of the different decorations. The origin of each Order, the rules and regulations, and the duties incumbent upon its members, are all given at full length. The fact of the work being under the supervision of Sir Bernard Burke, and endorsed by his authority, gives it another recommendation to the public favour."—Sun

" This is indeed a splendid book. It is an uncommon combination of a library book of reference and a book for a boudoir, undoubtedly uniting beauty and utility. It will soon find its place in every library and drawing-room."—Globe.

TRAITS OF CHARACTER; BEING TWENTY-FIVE

YEARS' LITERARY AND PERSONAL RECOLLECTIONS. By A CONTEMPORARY. 2 vols. 21s.

"'The Authoress of these volumes, having been thrown into communication with celebrities of all ranks and professions during the last quarter of a century, has naturally thought that her reminiscences of their ways and manners would prove interesting to readers of the present day. Prominent among the subjects of her sketches are Lord Melbourne, the Duke of Wellington, Edward Irving, Thomas Moore, Edmund Kean, Mr. Spurgeon, Lady Blessington, and Mrs. Shelley. Of the great Duke she gives a very interesting description. We commend these agreeable volumes to the reader, assuring him that he will find ample entertainment for a leisure hour in contemplating these varied and life-like photographs."—Sun.

THE ENGLISH SPORTSMAN IN THE WESTERN

PRAIRIES. By the Hon. GRANTLEY BERKELEY. Royal 8vo. with numerous Illustrations. (In February, 1861).

PICTURES OF SPORTING LIFE AND CHARACTER.

By LORD WILLIAM LENNOX. 2 vols. with Illustrations. 21s.

" This book should be in the library of every gentleman, and of every one who delights in the sports of the field. It forms a complete treatise on sporting in every part of the World, and is full of pleasant gossip and anecdote. Racing, steeple chasing, hunting, driving, coursing, yatching, and fishing, cricket and pedestrianism, boating and curling, pigeon shooting, and the pursuit of game with the fowling-piece, all find an able exponent in Lord William Lennox."—Herald.

REALITIES OF PARIS LIFE. BY THE AUTHOR

OF " FLEMISH INTERIORS," &c. 3 vols. with Illustrations. 31s. 6d.

"' Realities of Paris Life' is a good addition to Paris books, and important as affording true and sober pictures of the Paris poor."—Athenæum.

STUDIES FROM LIFE.   BY THE AUTHOR OF

" JOHN HALIFAX, GENTLEMAN," " A WOMAN'S THOUGHTS ABOUT WOMEN," &c. 1 vol. 10s. 6d. elegantly bound.

"Studies from Life is altogether a charming volume, one which all women and most men, would be proud to possess."—*Chronicle.*

" Without being in the same degree elaborate, either in purpose or plot, as 'John Halifax,' these ' Studies from Life' may be pronounced to be equally as clever in construction and narration. It is one of the most charming features of Miss Muloch's works that they invariably tend to a practical and useful end. Her object is to improve the taste, refine the intellect, and touch the heart, and so to act upon all classes of her readers as to make them rise from the consideration of her books both wiser and better than they were before they began to read them. The ' Studies from Life' will add considerably to the author's well earned reputation."—*Messenger.*

POEMS. BY THE AUTHOR OF " JOHN HALIFAX,

GENTLEMAN," " A WOMAN'S THOUGHTS ABOUT WOMEN," &c. 1 vol. with Illustrations by BIRKET FOSTER.

"A volume of poems which will assuredly take its place with those of Goldsmith, Gray, and Cowper, on the favourite shelf of every Englishman's library. We discover in these poems all the firmness, vigour, and delicacy of touch which characterise the author's prose works, and in addition, an ineffable tenderness and grace, such as we find in few poetical compositions besides those of Tennyson."—*Illustrated News of the World.*

" We are well pleased with these poems by our popular novelist. They are the expression of genuine thoughts, feelings, and aspirations, and the expression is almost always graceful, musical and well-coloured. A high, pure tone of morality pervades each set of verses, and each strikes the reader as inspired by some real event, or condition of mind, and not by some idle fancy or fleeting sentiment."—*Spectator.*

A SAUNTER THROUGH THE WEST END.   BY

LEIGH HUNT. 1 vol. *(Just Ready).*

NOVELS AND NOVELISTS, FROM ELIZABETH TO

VICTORIA. By J. C. JEAFFRESON, Esq. 2 vols. with Portraits. 10

THE RIDES AND REVERIES OF MR. ÆSOP SMITH.

By MARTIN F. TUPPER, D.C.L., F.R.S., Author of "Proverbial Philosophy," " Stephen Langton," &c., 1 vol. post 8vo. 5s.

THE MAN OF THE PEOPLE.   BY WILLIAM

HOWITT. 3 vols. post 8vo.

" A remarkable book, which refers to eventful times and brings before us some important personages. It cannot fail to make a powerful impression on its readers."—*Sun*

A JOURNEY ON A PLANK FROM KIEV TO EAUX-

BONNES. By LADY CHARLOTTE PEPYS. 2 vols, 21s

" A very beautiful and touching work."—*Chronicle.*

EASTERN HOSPITALS AND ENGLISH NURSES

The Narrative of Twelve Months' Experience in the Hospitals of Koula and Scutari. By A LADY VOLUNTEER. Third and Cheaper Edition 1 vol. post 8vo. with Illustrations, 6s. bound.

"The story of the noble deeds done by Miss Nightingale and her devoted sisterhood will never be more effectively told than in this beautiful narrative."—*John Bull.*

KATHERINE AND HER SISTERS.

By the Author of "THE DISCIPLINE OF LIFE," &c., 3 vols.

We always look forward with gratification whenever we take up one of Lady Emily Ponsonby's novels. 'Katherine and her Sisters' is without exception one of the very best of modern times."—*Messenger*.

THE HOUSE ON THE MOOR.

By the Author of "MARGARET MAITLAND," 3 v.

"This story is very interesting and the interest deepens as the story proceeds."—*Athenæum*.

THE WORLD'S VERDICT.

By the Author of "MORALS OF MAY FAIR," "CREEDS," &c. 3 vols.

"A remarkably able novel, and intensely interesting."—*Post*.

MAGDALEN HAVERING.

By the Author of "THE VERNEYS," 3 v

TWELVE O'CLOCK.

By the Author of "GRANDMOTHER'S MONEY," &c. 1 vol.

"An amusing story, full of point and vigour. No reader will lay it down till he has finished it."—*Messenger*.

THE VALLEY OF A HUNDRED FIRES.

By the Author of "MARGARET AND HER BRIDESMAIDS," &c. 3 vols.

"If asked to classify 'The Valley of a Hundred Fires' we should give it a place between 'John Halifax' and 'The Caxtons,'"—*Herald*.

HIGH PLACES.

By G. T. LOWTH Esq. 3 vols.

"A novel which contains interesting incidents, capitally drawn characters, and vivid pictures of life and society of the present day."—*Post*.

MONEY.

By COLIN KENNAQUHOM. 3 vols.

"A clever novel. It can hardly fail to amuse all readers."—*Spectator*.

DAUNTON MANOR HOUSE.

2 vols.

MY SHARE OF THE WORLD.

By FRANCES BROWNE, 3 vols.

THE DAILY GOVERNESS.

By the Author of "COUSIN GEOFFREY," &c. 3 vols.

BOND AND FREE.

By the Author of "CASTE," 3 vols.

"A clever and interesting novel. It has great power, and the story is well sustained."—*Literary Gazette*.

MAINSTONE'S HOUSE-KEEPER.

By SILVERPEN. 3 vols.

"The work of a very clever and able writer."—*Literary Gazette*.

THE CRAVENS OF BEECH HALL.

By MRS. F. GUISE. 2 vols.

GRANDMOTHER'S MONEY.

By the Author of "WILDFLOWER," 3 vols.

"A good novel. The most interesting of the Author's productions."—*Athenæum*.

CARSTONE RECTORY.

By GEORGE GRAHAM. 3 vols

"A brilliant novel."—*Sun*.

THE ROAD TO HONOUR.

"A very interesting story."—*Sun*.

NIGHT AND DAY.

By the Hon. C. S. SAVILLE. 3. vols.

ONLY A WOMAN.

By CAPTAIN L. WRAXALL. 3 vols.

LORD FITZWARINE.

By SCRUTATOR.

Author of "THE MASTER OF THE HOUNDS," &c. 2 vols., with Illustrations.

STEPHAN LANGTON.

By MARTIN. F. TUPPER. D.C.L. F.R.S. Author of "PROVERBIAL PHILOSOPHY." &c., 2 vols. with fine engravings. 10s.

THE CURATES OF RIVERSDALE.

Recollections in the Life of a Clergyman.

HURST AND BLACKETT'S STANDARD LIBRARY

OF CHEAP EDITIONS OF

POPULAR MODERN WORKS

Each in a single volume, elegantly printed, bound, and illustrated, price 5s.
A volume to appear every two months. The following are now ready.

VOL. I.—SAM SLICK'S NATURE AND HUMAN NATURE.

ILLUSTRATED BY LEECH.

" The first volume of Messrs. Hurst and Blackett's Standard Library of Cheap Editions
of Popular Modern Works forms a very good beginning to what will doubtless be a very
successful undertaking. 'Nature and Human Nature' is one of the best of Sam Slick's
witty and humorous productions, and well entitled to the large circulation which it
cannot fail to obtain in its present convenient and cheap shape. The volume combines
with the great recommendations of a clear, bold type, and good paper, the lesser, but
still attractive merits, of being well illustrated and elegantly bound."—*Morning Post.*

"This new and cheap edition of Sam Slick's popular work will be an acquisition to
all lovers of wit and humour. Mr. Justice Haliburton's writings are so well known to
the English public that no commendation is needed. The volume is very handsomely
bound and illustrated, and the paper and type are excellent. It is in every way suited
for a library edition, and as the names of Messrs. Hurst and Blackett, warrant the
character of the works to be produced in their Standard Library, we have no doubt the
project will be eminently successful."—*Sun.*

VOL. II.—JOHN HALIFAX, GENTLEMAN.

" This is a very good and a very interesting work. It is designed to trace the career
from boyhood to age of a perfect man—a Christian gentleman, and it abounds in incident
both well and highly wrought. Throughout it is conceived in a high spirit, and written
with great ability, better than any former work, we think, of its deservedly successful
author. This cheap and handsome new edition is worthy to pass freely from hand to hand,
as a gift book in many households."—*Examiner.*

"The new and cheaper edition of this interesting work will doubtless meet with great
success. John Halifax, the hero of this most beautiful story, is no ordinary hero, and this,
his history, is no ordinary book. It is a full-length portrait of a true gentleman, one of
nature's own nobility. It is also the history of a home and a thoroughly English one.
The work abounds in incident, and many of the scenes are full of graphic power and true
pathos. It is a book that few will read without becoming wiser and better."—*Scotsman*

VOL. III.—THE CRESCENT AND THE CROSS.

BY ELIOT WARBURTON.

"Independent of its value as an original narrative, and its useful and interesting
information, this work is remarkable for the colouring power and play of fancy with
which its descriptions are enlivened. Among its greatest and most lasting charms is its
reverent and serious spirit."—*Quarterly Review*

"A book calculated to prove more practically useful was never penned than 'The
Crescent and the Cross'—a work which surpasses all others in its homage for the sublime and its love for the beautiful in those famous regions consecrated to everlasting
immortality in the annals of the prophets, and which no other writer has ever depicted
with a pencil at once so reverent and so picturesque."—*Sun.*

VOL. IV.—NATHALIE. BY JULIA KAVANAGH.

"'Nathalie' is Miss Kavanagh's best imaginative effort. Its manner is gracious and
attractive. Its matter is good. A sentiment, a tenderness, are commanded by her which
are as individual as they are elegant. We should not soon come to an end were we to
specify all the delicate touches and attractive pictures which place 'Nathalie' high among
books of its class."—*Athenæum.*

"A tale of untiring interest, full of deep touches of human nature. We have no hesitation in predicting for this delightful tale a lasting popularity, and a place in the foremost
ranks of that most instructive kind of fiction—the moral novel."—*John Bull.*

"A more judicious selection than 'Nathalie' could not have been made for Messrs.
Hurst and Blackett's Standard Library. The series as it advances realises our first impression, that it will be one of lasting celebrity."—*Literary Gazette.*

[CONTINUED ON NEXT PAGE.]

HURST AND BLACKETT'S STANDARD LIBRARY
OF CHEAP EDITIONS.

Each in a single volume, elegantly printed, bound, and illustrated, price 5s.

VOL. V.—A WOMAN'S THOUGHTS ABOUT WOMEN.

BY THE AUTHOR OF "JOHN HALIFAX, GENTLEMAN."

"A book of sound counsel. It is one of the most sensible works of its kind, well-written, true-hearted, and altogether practical. Whoever wishes to give advice to a young lady may thank the author for means of doing so."—*Examiner.*

"The author of 'John Halifax' will retain and extend her hold upon the reading and reasonable public by the merits of her present work, which bears the stamp of good sense nd genial feeling."—*Guardian.*

"These thoughts are good and humane. They are thoughts we would wish women to think."—*Athenæum*

"This really valuable volume ought to be in every young woman's hand. It will teach her how to think and how to act."—*Literary Gazette.*

VOL. VI.—ADAM GRAEME, OF MOSSGRAY.

BY THE AUTHOR OF "MRS. MARGARET MAITLAND."

"'Adam Graeme' is a story awakening genuine emotions of interest and delight by its admirable pictures of Scottish life and scenery. The plot is cleverly complicated, and there is great vitality in the dialogue, and remarkable brilliancy in the descriptive passages, as who that has read 'Margaret Maitland' would not be prepared to expect? But the story has a 'mightier magnet still,' in the healthy tone which pervades it, in its feminine delicacy of thought and diction, and in the truly womanly tenderness of its sentiments. The eloquent author sets before us the essential attributes of Christian virtue, their deep and silent workings in the heart, and their beautiful manifestations in the life, with a delicacy, a power, and a truth which can hardly be surpassed."—*Morning Post.*

VOL. VII.—SAM SLICK'S WISE SAWS
AND MODERN INSTANCES.

"The best of all Judge allburton's admirable works.,"—*Standard.*

"'The humour of Sam Slick is inexhaustible. He is ever and everywhere a welcome visitor; smiles greet his approach, and wit and wisdom hang upon his tongue. The present production is remarkable alike for its racy humour, its sound philosophy, the felicity of its illustrations, and the delicacy of its satire. We promise our readers a great treat from the perusal of these 'Wise Saws and Modern Instances,' which contain a world of practical wisdom, and a treasury of the richest fun."—*Post.*

VOL. VIII.—CARDINAL WISEMAN'S RECOLLECTIONS
OF THE LAST FOUR POPES.

"A picturesque book on Rome and its ecclesiastical sovereigns, by an eloquent Roman Catholic. Cardinal Wiseman has here treated a special subject with so much generality and geniality, that his recollections will excite no ill-feeling in those who are most conscientiously opposed to every idea of human infallibility represented in Papal domination."—*Athenæum.*

"In the description of the scenes, the ceremonies, the ecclesiastical society, the manners and habits of Sacerdotal Rome, this work is unrivalled. It is full of anecdotes. We could fill columns with amusing extracts."—*Chronicle.*

VOL. IX.—A LIFE FOR A LIFE.

BY THE AUTHOR OF "JOHN HALIFAX, GENTLEMAN."

"We are always glad to welcome Miss Muloch. She writes from her own convictions, and she has the power not only to conceive clearly what it is that she wishes to say, but to express it in language effective and vigorous. In 'A Life for a Life' she is fortunate in a good subject, and she has produced a work of strong effect. The reader having read the book through for the story, will be apt (if he be of our persuasion) to return and read again many pages and passages with greater pleasure than on a first perusal. The whole book is replete with a graceful, tender delicacy; and in addition to its other merits, it is written in good careful English."—*Athenæum.*

[CONTINUED ON NEXT PAGE.]

HURST AND BLACKETT'S STANDARD LIBRARY
OF CHEAP EDITIONS.

Each in a single volume, elegantly printed, bound, and illustrated, price 5s.

(CONTINUED).

VOL. X.—THE OLD COURT SUBURB. BY LEIGH HUNT.

"A delightful book, of which the charm begins at the first line on the first page, for full of quaint and pleasant memories is the phrase that is its title, 'The Old Court Suburb.' Very full too, both of quaint and pleasant memories is the line that designates the author. It is the name of the most cheerful of chroniclers, the best of remembrancers of good things, the most polished and entertaining of educated gossips 'The Old Court Suburb' is a work that will be welcome to all readers, and most welcome to those who have a love for the best kinds of reading."—*Examiner*.

VOL. XI.—MARGARET AND HER BRIDESMAIDS.

"We may save ourselves the trouble of giving any lengthened review of this work, for we recommend all who are in search of a fascinating novel to read it for themselves. They will find it well worth their while. There are a freshness and originality about it quite charming, and there is a certain nobleness in the treatment both of sentiment and incident which is not often found."—*Athenæum*.

VOL. XII.—THE OLD JUDGE. BY SAM SLICK.

"The present work of Judge Haliburton is quite equal to the first. Every page is alive with rapid, fresh sketches of character, droll, quaint, racy sayings, good-humoured practical jokes, and capitally-told anecdotes."—*Chronicle*.

"These popular sketches, in which the Author of 'Sam Slick' paints Nova Scotian life, form the 12th Volume of Messrs Hurst and Blackett's Standard Library of Modern Works. The publications included in this Library have all been of good quality; many give information while they entertain, and of that class the book before us is a specimen. The manner in which the Cheap Editions forming the series is produced deserves especial mention. The paper and print are unexceptional; there is a steel engraving in each volume, and the outsides of them will satisfy the purchaser who likes to see a regiment of books in handsome uniform."—*Examiner*.

VOL. XIII.—DARIEN. BY ELIOT WARBURTON.

'This last production, from the pen of the author of 'The Crescent and the Cross,' has the same elements of a very wide popularity. It will please its thousands."—*Globe*.

"This work will be read with peculiar interest as the last contribution to the literature of his country of a man endowed with no ordinary gifts of intellect. We have seldom met with any work in which the realities of history and the poetry of fiction were more happily interwoven."—*Illustrated News*

VOL. XIV.—FAMILY ROMANCE; OR, DOMESTIC ANNALS
OF THE ARISTOCRACY.
BY SIR BERNARD BURKE, ULSTER KING OF ARMS.

"It were impossible to praise too highly as a work of amusement this most interesting book, whether we should have regard to its excellent plan or its not less excellent execution. It ought to be found on every drawing-room table. Here you have nearly fifty captivating romances with the pith of all their interest preserved in undiminished poignancy, and any one may be read in half an hour. It is not the least of their merits that the romances are founded on fact—or what, at least, has been handed down for truth by long tradition—and the romance of reality far exceeds the romance of fiction."—*Standard*.

VOL. XV.—THE LAIRD OF NORLAW.
BY THE AUTHOR OF "MARGARET MAITLAND."

"The author of this delightful work is favourably known to the reading public through several other books of the same class, but the present is, in our judgment, by far the best and most finished production of them all. Scottish life and character, in connection with the fortunes of the house of Norlaw, are here delineated with truly artistic skill. The plot of the tale is simple, but the incidents with which it is interwoven are highly wrought and dramatic in their effect, and altogether there is a fascination about the work which holds the attention spell-bound from the first page to the last."—*Herald*.

www.ingramcontent.com/pod-product-compliance
Lightning Source LLC
Chambersburg PA
CBHW032007110726
47901CB00004B/1002